CONTROL AND DYNAMIC SYSTEMS

Advances in Theory and Applications

Volume 25

ACADEMIC PRESS RAPID MANUSCRIPT REPRODUCTION

CONTROL AND DYNAMIC SYSTEMS

ADVANCES IN THEORY AND APPLICATIONS

Edited by
C. T. LEONDES

School of Engineering and Applied Science
University of California
Los Angeles, California

VOLUME 25: SYSTEM IDENTIFICATION AND ADAPTIVE CONTROL
Part 1 of 3

1987

ACADEMIC PRESS, INC.
Harcourt Brace Jovanovich, Publishers
Orlando San Diego New York Austin
Boston London Sydney Tokyo Toronto

ACADEMIC PRESS, INC.
Orlando, Florida 32887

United Kingdom Edition published by
ACADEMIC PRESS INC. (LONDON) LTD.
24–28 Oval Road, London NW1 7DX

LIBRARY OF CONGRESS CATALOG CARD NUMBER: 64-8027

ISBN 0–12–012725–3 (alk. paper)

PRINTED IN THE UNITED STATES OF AMERICA

87 88 89 90 9 8 7 6 5 4 3 2 1

CONTENTS

Discrete Averaging Principles and Robust Adaptive Identification

Robert R. Bitmead and C. Richard Johnson Jr.

Techniques for Adaptive State Estimation through the Utilization of Robust Smoothing

F. Dale Groutage, Raymond G. Jacquot, and R. Lynn Kirlin

Coordinate Selection Issues in the Order Reduction of Linear Systems

Andrei L. Doran

PREFACE

System identification adaptive control has matured to such an extent over the past 25 to 30 years that it is now quite appropriate to devote a volume of *Control and Dynamic Systems* to this subject. However, the proliferation of published material and new research in this field could not be encompassed in one volume; thus, this volume is the first of a trilogy to be devoted to System Parameter Identification and Adaptive Control.

The first contribution in this volume, Uncertainty Management Techniques in Adaptive Control, by H. V. Panossian, develops rather powerful and highly useful techniques for effective adaptive control where variations of the process or system parameters are described by a stochastic model. The uncertainties are taken into account in the derivation of the control algorithm in a general procedure called stochastic adaptive control systems. The well-known Russian research worker Feldbaum is perhaps most closely identified with the techniques of dual control, in which, for problems of system identification and adaptive control, it is desirable to balance the two conflicting roles of the system input, namely, learning and control.

In the second article, Multicriteria Optimization in Adaptive and Stochastic Control, by N. T. Koussoulas, a significant advance in dual control—a multicriteria optimization concept for balancing control and learning—is presented. The need for estimating the parameters of an autoregressive moving-average (ARMA) process arises in many applications in the areas of signal processing, spectral analysis, and system identification.

Instrumental Variable Methods for ARMA Models, by Stoica, Friedlander, and Söderström, is a notable contribution to this volume; the authors are preeminently identified with this subject internationally because of the significance of their contributions to this major area.

In the broad areas of system identification and adaptive control G. Goodwin and R. Middleton are certainly among the most visible contributors. Therefore, their contribution, Continuous and Discrete Adaptive Control is a most welcome and significant part of this volume. In this article they present robust algorithms for adaptive control laws. Their techniques utilize a two-stage approach for the design of adaptive controllers, and continuous and discrete systems are presented within a unified framework.

As noted by I. Bar-Kana in Adaptive Control: A Simplified Approach, the two main approaches in this subject can be classified as self-tuning regulators and model reference adaptive control. In the algorithms developed by Bar-Kana boundedness of all values involved in the presence of any bounded input or output disturbances is guaranteed. Additionally, in Bar-Kana's methods, the output tracking error can be controlled and arbitrarily reduced.

The sixth contribution, Discrete Averaging Principles and Robust Adaptive Identification, by R. R. Bitmead and C. R. Johnson, Jr., develops broad design principles for the achievement of robustness in adaptive identification methods. As such, this contribution constitutes an essential part of this trilogy.

The original developments of system state estimation were based on rather specific system equations, system input characteristics, knowledge of covariance matrices, etc. However, as the theory of stochastic filtering developed, it rapidly became apparent that extensions were needed in the theory in which the characteristics of these parameters were less specific. Techniques for Adaptive State Estimation through the Utilization of Robust Smoothing, by F. D. Groutage, R. G. Jacquot, and R. L. Kirlin, presents an excellent review of this area and the techniques for dealing with the situation wherein all the factors in stochastic filtering are not well defined. Important new techniques with respect to robustness are also presented.

The final contribution, Coordinate Selection Issues in the Order Reduction of Linear Systems, by A. L. Doran, deals with the fundamental issue of methods for the desirable or necessary reduction of the complexity of the equations used to describe a physical system. Original important advances are presented in this article. Because of the growing complexity of systems and its implication for the broad problems of system parameter identification and adaptive control, this is a particularly suitable contribution with which to conclude this first volume in the trilogy.

The authors are all to be commended for their superb contributions to this volume, which will most certainly be a significant reference source for workers on the international scene for many years to come.

Uncertainty Management Techniques in Adaptive Control

HAGOP V. PANOSSIAN

HR Textron Inc.
25200 West Rye Canyon Road
Valencia, California 91355

I. INTRODUCTION

A. *BACKGROUND*

Adaptive control methodology originated in the course of searching for high-performance self-tuning or self-organizing systems. The basic concept in adaptive control systems is the ability of the controllers to modify their behavior relative to the performance of the closed-loop system. The underlying functions for most adaptive regulators encompass identification of unknown parameters, decision of an optimal control strategy, and on-line modification of the parameters of the controller [1].

The class of adaptive control systems in which variations of the process parameters are described by a stochastic model and the uncertainties are taken into account in the derivation of the control algorithm are called stochastic adaptive control systems. The different ways of handling the above-mentioned types of problems and the philosophy involved in the derivations are treated in [2]. Stochastic adaptive control problems in which the Bayesian approach is taken can be found in [3].

In physical processes, the problem of engineering control system design, almost invariably, involves the on-line feedback control of uncertain, usually nonlinear systems. In a wide variety of situations, however, a control problem can be represented by a system of linear ordinary difference (or differential) equations characterized by a finite number of random parameters. The performance functional that is most commonly utilized is the quadratic cost functional. Thus linear quadratic Gaussian (LQG) control problems are the most commonly considered control systems [4].

Optimal solutions to stochastic adaptive control of linear dynamic systems with constant or time-varying uncertain parameters can be obtained, in principle, using the stochastic dynamic programming method [5]. However, since it is practically impossible to analytically solve the stochastic adaptive control problem (except when very special conditions prevail [6]), approximations are generally used. In all cases, the solution of the stochastic optimal control problem shows the effects of uncertainties in the performance of the control system [7]. In fact, in any stochastic optimal control system that contains randomly varying parameters, the control gain will be a function of the unconditional means and covariances of the uncertainties. The same dependence on the means and covariances will appear in the performance index of the system.

The above-mentioned dependence on the means and covariances of the control gain and the performance functional expressions of the stochastic control systems with multiplicative and additive noise leads to Riccati-like matrix difference (or differential) equations with random coefficients. The stability properties and the existence of solutions of the latter

Riccati-like equations are different from those of the constant parameter case and are closely connected with the behavior of the covariances of the uncertainties [8]. Systems with large uncertainties require a stringent stabilizability property for the existence of a solution [9,10,7].

It is a well-known fact that, under large parameter variations, representation of stochastic system uncertainties by multiplicative state and control-dependent noise is more realistic [7,11]. Furthermore, digital control systems with random sampling periods entail stochastic representations. Process control, economic systems, and systems controlled by humans in general are better modeled by discrete-time stochastic systems. Thus, discrete-time LQG systems with multiplicative noise encompass a wide scope of control problems, and the analysis of such systems enhances the usefulness of stochastic optimal control.

Design and control of any system encompasses consideration of (a) dynamics, (b) performance requirements, (c) disturbances/ uncertainties, and (d) availability of required observation and control authority. Common control synthesis techniques in present-day control design of structures entail (a) generation of a finite-element model and subsequently a state-space model of the structure, (b) development of a performance model, and (c) derivation of an appropriate disturbance model. However, development of a "correct" mathematical model of a system is still an art. Various errors, such as modeling errors, parametric errors, improperly modeled uncertainties, and errors due to linearization of nonlinear effects, create a challenging task of determining "best" models for a dynamic system. Structural control scientists make a big effort to compensate for most of the above-mentioned errors by developing very high-dimensional finite-element models and reducing the latter to lower order models through various reduction techniques involving system modes and frequencies. Large models are separated into controlled, suppressed, and residual parts, and only the controlled portion of the system model is finally treated, taking great pains in reducing the undesirable effects of model reduction by various modal and performance considerations. Yet, even with the best modeling and model reduction techniques, in the final analysis and implementation of control schemes, it is conceivable that better performance will be anticipated when uncertainties are modeled through stochastic multiplicative and additive noise terms. Optimal control strategies generated under all possible parameter variations will definitely create more robust control systems, under controllability and observability conditions, than those generated by the usual approaches. However, in most cases the engineer has to consider all of the constraints imposed by nature, economics, and the current state of technology to achieve a desired system strategy.

B. RANDOMNESS IN CONTROL SYSTEMS

Modeling and control design of systems entail three basic types of randomness, namely, uncertainties in modeling, uncertainties in the control system, and random disturbances that affect the performance of the system under consideration. The first of the above-mentioned stochastic phenomena is due to modeling of nonlinearities by approximation, model parameter errors, configuration changes, and internal as well as external disturbances. Uncertainties in the control system are also due to internal and external control-dependent noise as well as errors in positioning and activating of the controllers. The purely stochastic phenomena of the disturbances is even harder to account for, due to insufficient information (even statistical) regarding their characteristics. Of course, statistics are useful only because of the deficiency of humans in being able to generate exact knowledge of systems and phenomena. Theoretically, all systems would be deterministic if the control engineer knew everything regarding the control system and disturbances exactly. However, the study of uncertainty and randomness, based upon the firm foundation of the theory of probability, has made it possible to formulate very important problems realistically and to generate satisfactory solutions.

In order to properly evaluate a stochastic system one must understand the meaning of stochastic uncertainty in an abstract mathematical as well as practical sense. Different authors have different definitions for the concept of stochastic uncertainty [19], just as there are various definitions for the concept of information. For instance, suppose the main purpose of a stochastic control system is to decrease the uncertainty and increase the information through appropriate measurements. Then how should the question as to what is the best of two given measurement systems be answered? The one that decreases the variance or the one that decreases the entropy of some distribution? It then becomes necessary to separate control systems into Bayesian or other types of stochastic control systems. Moreover, separation in the above manner will enhance the quantitative meaning of the reduction of uncertainty or increasing information.

To clarify the previous conjecture let us consider a simple discrete-time control system given by the following equation:

$$x_{k+1} = \Phi(x_k, u_k, \theta_k), \tag{1}$$

where $x_k \in R^n$ is the state vector, $u_k \in R^m$ the control vector, and $\theta_k \in R^n$ the vector of stochastic disturbances and uncertainties. Let us address the usual control problem of determining the optimal (in some sense) control sequence $\{u_k\}_{k=1}^{N-1}$ that will minimize the mathematical expectation of a given performance functional J:

$$J = E\left[\sum_{k=1}^{n} L(x_k, u_k)\right]. \tag{2}$$

It is necessary to have a system of observations, represented here by [similar to the state Eq. (1)]

$$y_k = h(x_k, u_k, \gamma_k), \tag{3}$$

in order to come up with an optimal control strategy that will be able to minimize the given performance functional. By application of dynamic programming one can generate the best control law that will influence the future behavior of the control system by appropriately utilizing the supplied information from the measurement system. This last statement actually stresses the importance of selecting a measurement system, since once this selection is made the method of utilizing it is also selected through the optimal compensator of the system. Thus, in order to formulate the problem of stochastic uncertainty in a meaningful manner, two issues should be addressed: First, the objective of control system as a stochastic problem; second, the initial data regarding the mathematical model of the control system with the appropriate and consistent probability distribution. Only when the above-mentioned problems are settled can one ensure the required performance of the control system. In situations of high performance requirements, the issue of initial data regarding modeling of the control system is far more serious and difficult to handle. It is this problem that will be addressed herein.

C. *OCCURRENCE AND APPLICABILITY OF STOCHASTIC SYSTEMS WITH MULTIPLICATIVE AND ADDITIVE NOISE*

Even though the study of random algebraic equations have been of independent theoretical interest (in that their analysis leads to stochastic generalizations of classical results), many appplied problems in engineering, statistics, and mathematical physics lead to the synthesis of algebraic equations with random coefficients. An algebraic equation with random variable coefficients will ensue if, for instance, the coefficients are subject to random errors. This might be the case when coefficients of systems of equations are computed from experimental data, or when numerical considerations render it necessary that truncated values of the computed coefficients be used. Algebraic equations whose coefficients are random variables also arise in the spectral theory of matrices with random elements, with subsequent applications in mathematical physics and multivariate statistical analysis. In addition, there are a few points of intersection between the theories of random algebraic and random operator equations [13]. As is well established, the solution of a homogeneous difference (or differential) equation of nth order with constant coefficients is expressed as a function of the roots of the characteristic polynomial associated with the difference (or differential) operator. It is impossible to obtain closed-form expressions for the roots of the associated random characteristic polynomials in terms of the coefficients of the operator and to determine the distribution of the solution of the random equation.

The study of random power series was initiated by Borel and Steinhaus. Presently there exists an appreciable quantity of literature related to probabilistic methods in analysis, and the theory of random power series and related subjects are active fields of research. Furthermore, random polynomials can be considered within the frame work of probabilistic functional analysis.

Random matrices can also be defined as generalized random variables with values in the Banach algebra of $n \times n$ matrices. Furthermore, random matrices form an important class of matrices, and the analysis of their properties often leads to generalizations of known results in classical matrix theory. In recent times, however, random matrices have found numerous applications in mathematical statistics, physics, and engineering, where finite-dimensional approximations of random Hamiltonian operators are considered. Problems that are also treated include applied engineering problems related to systems of random linear equations. Products of random matrices have also led to some interesting classes of limit theorems in abstract probability theory [14].

Dealing with random matrices often leads to problems concerning random determinants, random characteristic polynomials, random eigenvalues, and their distribution, etc. A systematic analysis of the distribution of the roots of random characteristic polynomials would be of great practical importance and of great use in the study of systems of random algebraic, difference, and differential equations, stochastic stability theory, and eigenvalue problems for random integral equations.

The underlying objective in control theory is, of course, the mathematical modeling and control design of complex dynamic systems. In designing a control system, the engineer has to consider all constraints imposed by nature, economics, and the current state of technology to achieve a desired system strategy. In very general terms, the basic function of a control system is to maintain a given set of variables within definite bounds. In operating a physical system, perturbations (which cannot be exactly predicted and which subject the state of the system in question to strains and stresses) create the necessity of investigating control processes with the aid of stochastic models. Thus, the presence of change phenomena imposes basic constraints upon system performance.

Various concepts of stochastic stability arise as a natural consequence of the study of the qualitative behavior of a system that is subject to random disturbances. A priori admissible controls for applications at hand are determined by establishing effective stability criteria. On the other hand, the criterion chosen often leads to quantitative performance indices on the basis of which admissible control strategies can be meaningfully compared. The latter, in turn, leads to the well-known problem of optimization, that is, determination of an optimal control function within an admissible class of controls.

The stochastic dynamic model, described in the following section, provides a quite general framework for many optimization problems which arise naturally in the application of control theory. There are also, of course, limitations of such models. For instance, the fact that applied engineering control systems are nonlinear makes our model only an approximate representation of dynamic systems. However, the particular form chosen, especially the randomness of the state and control matrices, encompasses a very wide range of dynamic systems.

Due to their intrinsic importance and relative tractability, linear stochastic discrete systems occupy a central place in control problems. Optimal design of linear control systems with Wiener's [15] celebrated monograph on linear least-squares smoothing (filtering) of stationary time series opened new horizons. The developments following the above-mentioned monograph led to a high degree of achievement in the theory of control systems described by difference or differential equations.

As can easily be seen from our model, the disturbances present are of three types: control-dependent noise, state-dependent noise, and purely additive noise. Control-dependent noise could arise, for instance, as a result of fluctuations in system structure or in energy content. Also, the error committed while performing a control test can be modeled by a linear system with multiplicative noise in the control matrix [16]. State-dependent noise can be considered as any internal dynamic disturbance, which may be due to perturbations not explicitly included in the model, for instance, those in aerospace systems with high performance requirements [17]. Additive noise could arise as an environmental disturbance which acts together with control on the basic dynamics. Furthermore, an additional cause for multiplicative noise in dynamic system equations is modeling of system parameters as white noise processes. From our studies it is found that large state-dependent noise has a destabilizing effect on dynamics and tends to increase the magnitude of the optimal gain. Control-dependent noise, on the other hand, tends to increase error by diminishing the useful effect of control. Stabilization, especially by linear feedback, may not be possible for large state-dependent noise even if the controller is noise free.

II. OPTIMAL CONTROL OF STOCHASTIC LINEAR DISCRETE SYSTEMS WITH PERFECT MEASUREMENTS

Existence and uniqueness of an optimal control for stochastic linear discrete-time systems have been studied by various authors [9,10]. It is well known [18] that observable and stabilizable systems have unique solutions with asymptotically stable control systems in the steady-state situation. However, conditions for a solution of stochastic systems require stringent constraints on the statistical properties of system matrices.

A. *PROBLEM FORMULATION*

The problem is first presented in a general infinite-dimensional setting for completeness, and then the particular problem to be addressed is formulated. The notations utilized are similar to those used in [9] and [19].

1. *Optimal Control of Stochastic Systems in Hilbert Space*

Let H be a Hilbert space and $x(k) \in H$ be the state of the system considered at time k, where $k \in I = \{0, 1, \ldots, N\}$. Also, let $\bar{U}$ be the Hilbert space of controls and let $\Phi : H \to H$, $D : \bar{U} \to H$, $R : \bar{U} \to \bar{U}$, $Q : H \to H$, and $C : H_\eta \to H$ be bounded linear and continuous operators, where H_η is a separable Hilbert space and R an invertible positive-definite operator. Furthermore, $A(\cdot, \cdot)$, $B(\cdot, \cdot)$ are continuous bilinear transformations such that $A : H \times H_\zeta \to H$ and $B : H_\xi \to H$, where H_ζ, H_ξ are separable Hilbert spaces.

Consider the stochastic control system given by

$$x(k+1) = \Phi x(k) + DU(k) + A(x(k), \zeta(k)) + B(U(k), \xi(k)) + C_\eta(k), \tag{4}$$

where $x(0)$, $\zeta(0)$, $\zeta(1)$, ..., $\xi(0)$, $\xi(1)$, ..., $\eta(0)$, $\eta(1)$, ... are mutually independent zero-mean random variables with finite second moments taking their values in appropriate Hilbert spaces. $\zeta(0)$, $\zeta(1)$, ..., $\xi(0)$, $\xi(1)$, ..., $\eta(0)$, $\eta(1)$, ... are identically distributed and Σ_ζ, Σ_ξ, and Σ_η and their respective covariance operators are defined by $\Sigma_\zeta : H_\zeta \to H_\zeta$, $\Sigma_\xi : H_\xi \to H_\xi$, and $\Sigma_\eta : H_\eta \to H_\eta$. The covariance operators are such that

$$\Sigma_\zeta(x, y) = E(\zeta, x)(\zeta, y) \quad \text{for all} \quad x, y \in H$$

and

$$\operatorname{tr} \Sigma_\zeta = \sum_{k=1}^{\infty} (\Sigma_\zeta e_k, e_k) = E|\zeta^2|,$$

where $\{e_k\}_{k=1}^{\infty}$ is an orthonormal basis in H_ζ and Σ_ζ a nuclear self-adjoint operator. It is assumed that $E[\zeta_k] = E[\xi_k] = E[\eta_k] = 0$ for $k = 0, 1, \ldots$, where $E[\cdot]$ denotes the statistical expectation operator.

The most convenient cost functional for the above system is normally the quadratic one given by

$$J(U) = E\left[(Fx_N, x_N) + \sum_{k=0}^{N-1} \{(Qx(k), x(k)) + (RU(k), U(k))\}\right], \tag{5}$$

where Q is a bounded, self-adjoint, positive-semidfinite operator. That is, for some b_0, b_1, b_2, > 0, $0 \le (x_i, Qx_i) \le b_0 \|x_i\|$

for all $x_i \in H$. Also, $b_1 \|U_i\|^2 < (U_i, RU_i) \leq b_2 \|U_i\|^2$ for all $U_i \in \overline{U}$, where $(\cdot, \cdot)$ and $\|\cdot\|$ denote the inner product and the norm on H, respectively.

The stochastic optimal control problem is then to determine the optimal control sequence $\{U_i^*\}_{i \in I}$ with $U_i^* \in \overline{U}$ such that the functional J(U) in Eq. (17) is minimized.

Theorem 1 [7]

Suppose there is an operator W such that the spectral radius of G_W, $\rho(G_W) < 1$, where

$$G_W(k) = \sum_{n=1}^{\infty} \lambda_n A_n^* K_n + W^* \left(\sum_{n=1}^{\infty} \mu_n B_n^* K B_n \right) W + (\Phi - DW)^* K (\Phi - DW) \tag{6}$$

and where $(^*)$ is complex transposition, λ_n, μ_n, $n = 1, 2, \ldots,$ are the eigenvalues of the operators Σ_ζ and Σ_ξ, respectively, and $A_n(\cdot) = A(\cdot, e_n)$ and $B_n(\cdot) = B(\cdot, e_n)$, $n = 1, 2, \ldots$. Then (i) for any K positive semidefinite, the sequence $\{R^n(K); n = 1, 2, \ldots\}$, where

$$R^n(K) = \left\{ Q^* Q + \sum_{n=1}^{\infty} \lambda_n A_n^* K A_n + \Phi^* K \left[I + D \left(R + \sum_{n=1}^{\infty} \mu_m B_n^* K B_n \right)^{-1} D^* K \right]^{-1} \right\}^n \tag{7}$$

is bounded. (ii) If (i) is satisfied, then there exists at least one nonnegative solution to

$$R(K) = K, \tag{8}$$

which is nothing but the algebraic Riccati equation. (iii) If, for some n_0, $R^{n_0}(0)$ is invertible, then $\rho(G_{W_{\overline{K}}}) < 1$, $\overline{K}$ is the unique solution of Eq. (8), and $|R^n(K) - \overline{K}| \to 0$ geometrically. (For a proof see [9].)

Theorem 2 [9]

Let $F_W(K) = Q^*Q + W^*RW + G_W(K)$; then (i) if the conditions of Theorem 1 hold and $K = F_W(K)$, $\Phi = -W$ is an admissible control law. In addition, the minimal cost is

$$J = \text{tr}\ C^*KC\Sigma_\eta. \tag{9}$$

(ii) If $\overline{K}$ is the solution of Eq. (8), then the minimal cost becomes

$$\overline{J} = \text{tr}\ C^*KCR. \tag{10}$$

2. *Optimal Control of Linear Discrete-Time Systems with Stochastic Parameters in the Finite-Dimensional Euclidean* n *Space*

Consider the stochastic system given by

$$x_{k+1} = \Phi_k x_k + \Psi_k U_k, \tag{11}$$

where $x_k \in R^n$ is the state vector (where R^n denotes the Euclidean n space with the usual inner product and norm, $U_k \in R^n$ the control vector, and $\Phi_k \simeq n \times n$, $\Psi_k \simeq n \times m$ the system matrices. It is assumed here (following [10]) that Ψ_k are sequences of random matrices that have independent elements with stationary statistics.

Moreover, it is assumed that the initial state x_0 is deterministic. Now, suppose the controls are of the linear feedback type:

$$U_k = -Gx_k, \tag{12}$$

where $G \simeq m \times n$. Then the closed-loop system will be

$$x_{k+1} = (\Phi_k - \Psi_k G)x_k. \tag{13}$$

Theorem 3 [10]

(i) The closed-loop system in Eq. (13) is stable in the mean if $E[(\Phi_k - \Psi_k G)]$ is stable and (ii) is stable in the mean-square sense if the real symmetric operator L defined by

$$LX = E[(\Phi - \Psi G)^T X(\Phi - \Psi G)] \tag{14}$$

for every real symmetric $n \times n$, stable matrix X, is stable (here stability refers to the condition that the spectral radius is less than unity). (iii) The closed-loop system is stabilizable in the mean if there exists a matrix G such that $E[\Phi_k - \Psi_k G]$ is stable and is stabilizable in the mean-square sense if there exists a G such that L is stable.

Consider the dynamic system given in Eq. (11) with the quadratic cost functional of Eq. (5) but now Q and R are $n \times n$ and $m \times m$ positive-semidefinite and positive-definite matrices, respectively.

Theorem 4 [10]

Suppose the optimal control law is given by Eq. (12); then the optimal transient cost is given by

$$J_N(U_N, x_0) = x_0^T[(\Phi - \Psi G)^T x(\Phi - \Psi G) + G^T RG]^N F_M x_0 \quad \forall x_0 \tag{15}$$

and the optimal control gain matrix G is

$$G = -(E[\Gamma^T X\Gamma] + R)^{-1} E[\Gamma^T X\Phi]. \tag{16}$$

B. OPTIMAL CONTROL OF STOCHASTIC LINEAR DISCRETE-TIME SYSTEMS FOR THE PSEUDO-DETERMINISTIC CASE

In stochastic optimal control problems it is vital to furnish the required information for the control system. Hence, in the present section it will be assumed that all the states can be measured exactly. Moreover, it is assumed that the admissible controls are real valued and of the state-feedback type, that is, $u(k) = f(x(k), k)$, and that at time k they can only influence the states $x(i)$ at $i \geq k + 1$.

Consider a dynamical system represented by the following vector difference equations:

$$x(k+1) = \begin{bmatrix} x_1(k+1) \\ x_2(k+1) \end{bmatrix} = \begin{bmatrix} A(k) & 0_{n_1 \times n_2} \\ 0_{n_2 \times n_1} & \theta(k) \end{bmatrix} x(k) + \begin{bmatrix} B(k) \\ \Gamma(k) \end{bmatrix} u(k) + \begin{bmatrix} 0 \\ \xi_0(k) \end{bmatrix}, \quad k = 0, 1, \ldots, N. \tag{17}$$

In Eq. (17) x and u are the n- and m-dimensional state and control vectors respectively, θ and Γ are $n_2 \times n_2$- and $n_2 \times$ m-dimensional matrices of randomly varying Gaussian white parameters with statistics given by

$$E[\Theta(k)] = \overline{\Theta}(k), \quad E[(\Theta(k) - \Theta(k))(\Theta(j) - \Theta(j))^T] = \Sigma^{\Theta\Theta}\delta_{kj}$$

and

$$E[\Gamma(k)] = \overline{\Gamma}(k), \quad E[(\Gamma(k) - \overline{\Gamma}(k))(\Gamma(j) - \overline{\Gamma}(j))^T] = \Sigma^{\Gamma\Gamma}\phi_{kj},$$

and A and B are $n_1 \times n_1$- and $n_1 \times$ m-dimensional deterministic matrices, respectively. ξ_0 is an n_2-dimensional zero-mean white Gaussian noise vector independent of all the other random variables, $E\left[\xi_0(k)\xi^T(j)\right] = \Sigma^{\xi_0\xi_0}\delta_{kj}$, and $0_{n_1 \times n_2}$ and $0_{n_2 \times n_1}$ are null matrices of the indicated dimensions. Note that δ_{kj} is the Kronecker delta operator.

Equations like Eq. (17) occur in situations when the additive noise vector of a linear dynamic system is non-white ("colored") noise. The noise vector ξ is then modeled as

$$\xi(k+1) = \theta(k)\xi(k) + \xi_0(k), \tag{18}$$

where $\xi_0(k)$ is white Gaussian. In such cases, where a usual, deterministic linear system is augmented with Eq. (18), the result will be Eq. (17).

The stochastic optimal control problem is now to determine a control sequence $\{u_i\}_{i=0}^{N-1}$ such that the functional J(U) in Eq. (5) is minimized.

For simplicity, Eq. (17) can be rewritten as

$$x(k + 1) = \Phi(k)x(k) + \Psi(k)u(k) + \xi(k). \tag{19}$$

The appropriate dimensions of x, $\overline{\Psi}$, and ξ can be deducted from Eq. (17). The information set, in this case, will be

$$I(k) = \{x(0), x(1), \ldots, x(k), u(0), u(1), \ldots, u(k - 1)\}. \tag{20}$$

Theorem 5

The optimal control for the system given by Eq. (19) minimizing J(U) in Eq. (5) is

$$U(k) = -G(k)x(k), \tag{21}$$

where

$$G(k) = [R + E[\Psi^T(k)K(k + 1)\Psi(k)]^{-1}] \times E[\Psi^T(k)K(k + 1)\Phi(k)], \tag{22}$$

and where

$$K(k) = Q + E[\Phi^T(k)K(k + 1)\Phi(k)] - E[\Phi^T(k)K(k + 1)\Psi(k)] \times [R + E[\Psi^T(k)K(k + 1)\Psi(k)]]^{-1} \times E[\Psi^T(k)K(k + 1)\Phi(k)] \tag{23}$$

with

$$K(N) = F_N \tag{24}$$

is the Riccati-like matrix difference equation.

The proof of the above theorem is by dynamic programming and is straightforward. The reader is referred to [7] for details. Furthermore, evaluation of expressions like $E[\Phi^T(k)K(k + 1)\Phi(k)]$ is not an easy matter. Thus, we establish the following lemma for the above purpose.

Lemma 1

The expectation

$$E[\Phi^T(k)K(k+1)\Phi(k)] = \overline{\Phi}^T(k)K(k+1)\overline{\Phi}(k) + \mathrm{Tr}(\Sigma^{\Phi\Phi}K(k+1)), \qquad (25)$$

where $\Sigma^{\Phi\Phi}$ is the covariance matrix and $\overline{\Phi}$ is the mean value of Φ. Furthermore, $\mathrm{Tr}(\cdot)$ is the matrix trace operator given by

$$\mathrm{Tr}(\Sigma^{\Phi\Phi}K) = \begin{bmatrix} \mathrm{tr}(\Sigma^{\Phi_1\Phi_1}K) & \mathrm{tr}(\Sigma^{\Phi_1\Phi_2}K) & \cdots & \mathrm{tr}(\Sigma^{\Phi_1\Phi_n}K) \\ \vdots & & & \\ \mathrm{tr}(\Sigma^{\Phi_n\Phi_1}K) & & \cdots & \mathrm{tr}(\Sigma^{\Phi_n\Phi_n}K) \end{bmatrix}$$

where $\Sigma^{\Phi_i\Phi_j}$ indicates the covariance between the ith column and the jth. For a proof see [7].

C. UNIQUENESS OF OPTIMAL CONTROLLER

The extremal control of a stochastic multivariable discrete-time dynamical system is not necessarily the unique optimal control. The second-order partial derivative of the Hamiltonian function with respect to the controller u must be shown to be positive definite in order to ensure uniqueness [20,21].

The Hamiltonian functional for the present problem is given by (after some mathematical manipulation):

$$H = \frac{1}{2}\, x^T(k)K(k)x(k) + \sum_{i=k}^{N-1} \mathrm{Tr}(K(i+1)\Sigma^{\xi\xi}) + P^T(\Phi(k)x(k) + \Psi(k)u(k) + \xi(k)). \qquad (26)$$

Hence,

$$\partial^2 H/\partial^2 u = R(k) + E[\Psi^T(k)K(k+1)\Psi(k)]. \qquad (27)$$

Theorem 6

The solution K(k) to the Ricatti-like equation (23) with (24) is nonnegative definite and is unique for $N < \infty$. For a proof see [7].

Remark 1

The control gain G(k) is definitely a function of the covariances of the random white parameters that comprise the system matrices. Thus, G(k) will be smaller in norm than the regular LQG gain when only $\Gamma(k)$ is uncertain in Eq. (17), and it will be larger in norm when only $\Theta(k)$ is uncertain, with a larger covariance $\Sigma^{\Theta\Theta}$. It is then clear that the certainty equivalence principle will not hold in the present situation. In the latter case (i.e., when the certainty equivalence principle holds), all the covariance matrices will drop out.

Remark 2

The stochastic optimal control in the above problem is without a posteriori learning. The parameters in the system, being random, render it necessary for the controller to *adapt* to the structural and dynamic variations due to nature or change.

D. ASYMPTOTIC BEHAVIOR OF THE SOLUTION OF THE RICCATI EQUATION

Important aspects of the Riccati matrix equations are treated in [22], [23], and [24].

Let us assume that the stochastic linear system given in Eq. (23) has wide-sense stationary statistics and that the state and control weighting matrices Q and K are constant. Then, the

Riccati-like matrix difference equation takes the following form:

$$
\begin{aligned}
K(k) = {} & Q + \bar{\Phi}^T K(K + 1)\bar{\Phi} + \mathrm{Tr}(\Sigma^{\Phi\Phi} K(k + 1)) \\
& - [\bar{\Phi}^T K(k + 1)\bar{\Psi} + \mathrm{Tr}(\Sigma^{\Phi\Psi} K(k + 1))] \\
& \times [R + \bar{\Psi}^T K(k + 1)\bar{\Psi} + \mathrm{Tr}(\Sigma^{\Psi\Psi} K(k + 1))]^{-1} \\
& \times [\bar{\Psi}^T K(k + 1)\bar{\Phi} + \mathrm{Tr}(\Sigma^{\Phi\Psi} K(k + 1))]
\end{aligned} \tag{28}
$$

with

$$
K(N) = 0. \tag{29}
$$

It is worth noting that Eq. (28) does *not* always have a steady-state solution "backward in time" because, unlike the constant parameter situation, here we have constant but unknown parameters. The solution for the infinite interval of the scalar form of Eq. (28) has been treated previously [25]. Furthermore, unlike the Riccati equation, Eq. (28) cannot be related to a coupled set of linear equations. Hence, only under some conditions can the existence of a steady-state solution be investigated.

It is not hard to show that there exist matrices T_1, T_2, and T_3 such that the following equality holds [26,27,28]:

$$
\bar{\Phi}^T K(k + 1)\bar{\Phi} + \mathrm{Tr}\ \Psi^{\Phi\Phi}(k + 1) = T_1 \bar{\Phi}^T K(k + 1)\bar{\Phi} T_2. \tag{30}
$$

Assuming the above is true, we can rewrite Eq. (28) in the following manner:

$$
\begin{aligned}
K(k) = {} & Q + T_1 \bar{\Phi}^T K(k + 1)\bar{\Phi} T_2 - T_2^T \bar{\Phi}^T K(k + 1)\bar{\Psi} \\
& \times \left[R + T_3 \bar{\Psi} K(k + 1)\bar{\Psi}\right]^{-1} \bar{\Psi}^T K(k + 1)\bar{\Phi} T_2.
\end{aligned} \tag{31}
$$

By adding and subtracting $T_2\Phi^T K(k+1)\Phi T_2$ to Eq. (31), we obtain

$$K(k) = Q + \left(T_2 - T_2^T\right)\overline{\Phi}^T K(k+1)\overline{\Phi}T_2 + T_2^T\overline{\Phi}^T \times \left\{K(k+1) - K(k+1)\overline{\Phi}R + T_3\overline{\Psi}^T K(k+1)\overline{\Psi}^{-1}\overline{\Phi}^T K(k+1)\right\}\overline{\Phi}T_2. \quad (32)$$

Now, consider the quantity inside the curly braces and define

$$M(k+1) = K(k+1) - K(k+1)\overline{\Psi} \times \left[R + T_3\overline{\Psi}^T K(k+1)\overline{\Psi}\right]^{-1}\overline{\Psi}^T K(k+1). \quad (33)$$

Remark 3

In [29] it is given that for matrices A, B, and C of appropriate dimensions it is true that

$$(A + BCB^T)^{-1} = A^{-1} - A^{-1}B(C^{-1} + B^T A^{-1}B)^{-1}B^T A^{-1}.$$

Hence, in Eq. (33) $M(k+1)$ can be rewritten as

$$M(k+1) = K^{-1}(k+1) + \overline{\Psi}^T\left[R + T_2(\Sigma^{\Psi\Psi}K(k+1))\right]\overline{\Psi}^{1-^{-1}}.$$

It is now clear why $M(k+1)$ need be positive definite. Note also that for $K(k+1)$ very large, we can approximate Eq. (32) by

$$K(k) = Q + \left(T_1 - T_2^T\right)\overline{\Phi}^T K(k+1)\overline{\Phi}T_2,$$

which clearly is a Lyapunov-type equation and can be treated as such.

Matrices of the form of Eq. (33) arise in the matrix Riccati equation of standard linear quadratic Gaussian problems, where the weighting matrices T_1, T_2, and T_3 are not necessarily unique [23]. If we assume that the pairs of matrices $(\overline{\Phi}, \overline{\Psi})$ and $(\overline{\Phi}, Q^{1/2})$ are controllable and observable, respectively, then it is established [23,25] that

$$M(k+1) = M^T(k+1) > 0 \quad (34)$$

and that there exists a bound L such that

$$L \geq M(k + 1) \quad \text{for all} \quad k. \tag{35}$$

Hence, it follows that*

$$T_2^T \bar{\Phi}^T M(k + 1) \Phi^T T_2 > 0. \tag{36}$$

Thus,

$$K(k + 1) \geq (T_1 - T_2)\bar{\Phi}^T K(k + 1)\bar{\Phi} + Q. \tag{37}$$

Obviously, if any eigenvalue of $(T_1 - T_2)\bar{\Phi}^T$ is greater than unity (i.e., if the spectral radius is greater than one), then K(k) grows without bound "backward in time." In other words, $\lim_{k\to\infty} K(k)$ does not exist. So, the optimal cost grows exponentially as

$$J^*(N) \geq D \exp \max_i |\lambda_i| N, \tag{38}$$

where D is a constant matrix and $\max |\lambda_i|$ denotes the magnitude of the maximum eigenvalue of $(T_1 - T_2)^{i}\bar{\Phi}^T$. In such a case, of course, only short-term controls can be implemented. From Eqs. (33) and (35) we have

$$K(k) \leq (T_1 - T_2)\bar{\Phi}^T K(k + 1)\Phi^T + Q + T_2^T \bar{\Phi}^T L \bar{\Phi} T_2. \tag{39}$$

Consequently, if $|\lambda_i| < 1$ for all i, the right-hand side of inequality (39) will be a constant bounded solution matrix and so will K(k). Thus, the limiting solution $\lim_{k\to\infty} K(k)$ is well defined. For more details of the above argument, see [30] and [9], where it is required that Ψ be $n \times n$ nonsingular.

Remark 4

For $\Sigma^{\Phi\Phi} = \Sigma^{\Phi\Psi} = 0$ the infinite time problem has a solution independent of λ_i [13], whereas, when the above covariances get

Here, the notation $A > 0$ stands for positive definiteness of the matrix A.

larger, no stability can be expected in K(k). Hence, if we define

$$\beta = \max_{i,j} |(T_1 - T_2)_{ij}| \tag{40}$$

as the maximum value of the elements of the matrix $(T_1 - T_2)$ then, if

$$\max_i |\lambda_i| < 1/\beta,$$

the solution of the steady state will exist and $1/\beta$ will give the radius of a shrinking disk which will contain all open-loop eigenvalues of $\overline{\Phi}$ that make the problem solvable.

E. CLOSED-LOOP STOCHASTIC STABILITY

Consider the closed-loop system given by

$$x(k + 1) = (\Phi(k) - \Psi(k)G(k))x(k) + \xi(k), \tag{41}$$

where G(k) is given by Eq. (22). It is shown in [7] that there exists a limiting gain matrix G that is constant. However, the steady-state control (or the limiting control) generates unbounded states due to situations of very large covariances of the stochastic parameters. Thus, since it is assumed in the present section that x(k) can be measured exactly [i.e., x(0) is given], the mean value of the states $E[x(k)] = \overline{x}(k)$ will propagate (in an open-loop sense) according to

$$\overline{x}(k + 1) = (\Phi(k) - \Psi(k)G\overline{x}(k), \quad x(0) = \overline{x}(0). \tag{42}$$

Now, the state error covariance matrix defined by

$$E[(x(k) - \overline{x}(k))(x((i) - x(i))^T] \triangleq \Sigma^{xx}(k)\delta_{ik} \tag{43}$$

will obey the following equation:

$$\Sigma^{xx}*k + 1) = E[(\Phi - \Psi G)x - (\Phi - \Psi G)\overline{x}] \times [(\Phi - \Psi G)x - (\Phi - \Psi G)\overline{x}]^T, \tag{44}$$

which, after some manipulation, yields

$$\Sigma^{xx}(k+1) = (\overline{\Phi} - \overline{\Psi}G)\Sigma^{xx}(k)(\overline{\Phi} - \overline{\Psi}G)^T + \text{Tr}\, \Sigma^{xx}(k)\Sigma^{DD} - (\overline{\Phi} - \overline{\Psi}G)\overline{xx}^T(\overline{\Phi} - \overline{\Psi}G)^T \tag{45}$$

with

$$\Sigma^{xx}(0) = 0, \tag{46}$$

A simpler form from Eq. (45) can now be acquired by applying the same kind of transformation that we applied in Section II,D. Hence, there exists a matrix M which will express Eq. (45) in the following concise manner:

$$\Sigma^{xx}(k+1) = M\Sigma^{xx}(k)M^T + M\overline{xx}^T. \tag{47}$$

It can be shown [32,33] that if any one of the eigenvalues of M is greater than unity, then the open-loop propagation of the variance of the state vector $\Sigma^{xx}(k)$ will be unstable. The consequence of the above fact is, essentially, that even though the steady-state control (or more properly the limiting control) is well defined with a constant gain matrix and the closed-loop system, Eq. (41) can be implemented, yet the variability of the states as measured by the variances gets out of hand.

We will further study the stochastic stability properties of linear discrete-time multivariable systems under feedback control. The mathematical formulation of the stochastic control problem has been treated by various authors [31,32,33,23, 10].

Stability results will be invariant whether additive noise is included in or excluded from the system dynamics. Hence, for convenience, we will consider the following:

$$X(k+1) = \Phi(k)x(k) + \Psi(k)U(k). \tag{48}$$

We will analyze the stabilizability of Eq. (48) under the previously given conditions and the feedback control law of

$$u(k) = G(k)x(k). \tag{49}$$

The closed-loop system will now propagate in the following manner:

$$x(k + 1) = [\Phi(k) + \Psi(k)G(k)]x(k) = D(k)x(k). \tag{50}$$

Now, if we make the assumption that $\Phi(k)$ and $\Psi(k)$ are uncorrelated in time, following [25], we can calculate the ratio of the second moment of the state vector with the second moment of the initial state as a measure of stability.

Thus, let us assume that $D(k)$ is sequentially independent. Then, define

$$S(k) \triangleq \frac{E[x^T(k + 1)x(k + 1)]}{E[x^T(0)x(0)]} = \frac{tr(E[x(k + 1)x^T(k + 1)])}{tr(E[x(0)x^T(0)])}$$

$$= tr((E[D(0)D^T(0)]\ E[D(1)D^T(1)]) \cdots (E[D(k)D^T(k)]). \tag{51}$$

Consider the case when $S(k)$ grows without bound. Clearly, then the system of Eq. (50) will be unstable in the mean-square sense. Now, since the feedback gain matrix $G(k)$ could be crucial to the value of $S(k)$, we will have to make a judicial choice of $G(k)$. Obviously, the best choice will be that which minimizes $S(k)$ in Eq. (51). Thus, by the usual approach, we have

$$\frac{\partial tr(E[D(k)D^T(k)])}{\partial G(k)} = 0. \tag{52}$$

After some mathematics, we obtain

$$tr(E[DD^T])\big|_G = N < \infty, \tag{52}$$

and hence

$$S^*(k) = N^k \tag{54}$$

is the minimum value of S(k) at k. Clearly,

$$\lim_{k\to\infty} S^*(k) < \infty \quad \text{iff} \quad N < 1. \tag{55}$$

The previous result is formally stated in the next theorem.

Theorem 7

The linear stochastic system of Eq. (48) is (on the average) stabilized under feedback control in a mean-square sense if the "uncertainty threshold" parameter N is less than unity.

Using the notation of [33], [34], and [25], we now analyze the stochastic linear system of Eq. (48) under feedback control of Eq. (49) for "almost sure" Lyapunov stability.

Definition 1 (Kozin)

The equilibrium solution x(k) = 0 of the system of Eq. (50) with $x(0) = x_0$ (a random variable) is almost surely stable (ass) if, given any $\epsilon \to 0$, there exists a $\delta(\epsilon) > 0$ such that

$$\lim_{\delta\to 0} P\{\sup_{x_0} < \delta \ \sup_{k>0} ||x(k; x_0)|| > \epsilon\} = 0, \tag{56}$$

where $||x_0||$ stands for the Euclidian norm in R^n, P is the probability, and sup is the supremum.

For discrete-time systems, the above definition can be rewritten as follows.

Definition 2 (Konstantinov)

The equilibrium solution x(k) = 0 of the system of Eq. (50) is ass if for some $\epsilon > 0$,

$$|| \lim_{x_0\to 0} || P\{\sup_{k\to 0} || x(k, x_0) || > \epsilon\} = 0. \tag{57}$$

It has been proven [34] that the equilibrium solution x(k) = 0 for the system of Eq. (50) is a for $K \geq 0$ under some restrictive conditions to be given below.

Theorem 8

The solution $x(k) = 0$ of the system of Eq. (50) is ass for $k \to 0$ if there exists a function $V(x, k) \in D_L$ [where D_L is the domain of definition of the generator $V(x, k)$ consisting of functions that are integrable with respect to the probability measure chosen], which satisfies the following conditions for $k \geq 0$:

(i) $V(x, k)$ is continuous for $x = 0$ and $V(0, k) = 0$;

(ii) $\inf V(x, x) > \alpha(\delta)$ for any $\delta > 0$, $||x|| < \delta$;

(iii) $L[V(x, k)] \leq 0$ in some neighborhood of $x = 0$.

An appropriate Lyapunov function for the problem analyzed in our case is

$$V(x, k) = x^T(k)x(k) = \operatorname{tr} x(k)x^T(k). \tag{58}$$

From condition (iii) in Theorem 8 we have

$$E[V(k + 1, x) - V(k, x)] \leq 0. \tag{59}$$

Upon application of Eq. (50) and the properties of the trace operator, we are led to

$$\operatorname{tr}(E[\Phi + \Psi G]) < 1. \tag{60}$$

Hence, almost every single sequence of vectors $\{x(k)\}$ will approach zero.

We will now show (following [25]) that the equilibrium solution $x(k) = 0$ of Eq. (50) is ass (in the sense of Lyapunov) if Eq. (60) holds.

Theorem 9

The equilibrium solution $x(k) = 0$ of Eq. (50) is almost surely Lyapunov stable if Eq. (60) holds true.

The above theorem indicates that the mean-square stability condition is stronger than the almost sure stability criterion, because the latter states that the equilibrium solution $x(k) = 0$

is stochastically stabilizable. It ensures the existence of a control that will drive the system to zero (except for random fluctuations). Its criterion is based upon the ensemble of the sample trajectories and takes for granted the finiteness and boundedness of x(k) if Eq. (55) is satisfied.

F. THE DISCOUNTED COST PROBLEM

The space where the optimal control for the infinite horizon problem is well defined can be extended into a larger space when the cost functional is weighted by a discount factor α^k, where $0 < \alpha < 1$ [35]. Discount factors are used extensively in economics to emphasize the short-term worth of a utility function as compared to the long-term worth [36].

In applied dynamic systems control, the discount factor has been used for infinite-time control problems [37]. In the infinite horizon problem, since the cost is infinite, it is usually normalized by the planning horizon N. Thus discounted cost functionals for discrete systems are expressed by

$$J_{av} = \lim_{N\to\infty} \frac{1}{N} E\left\{\sum_{k=1}^{N} x^T(k)Q(k)x(k) + u^T(k)R(k)u(k)\right\}. \tag{61}$$

It can be shown [37] that J_{av} can be closely approximated by

$$J_d = \sum_{k=0}^{\infty} \alpha^k [x^T(k)Q(k)x(k) + u^T(k)R(k)u(k)] \tag{62}$$

for $0 > \alpha > 1$.

The use of the discount factor α guarantees the finiteness of the cost for any stabilizing control. Furthermore, in the above situation, the initial performance is emphasized, and if detectability and stabilizability conditions are not met, the Riccati-like equation may have multiple nonnegative-definite solutions. In Eqs. (61) and (62), $Q > 0$ and $R > 0$ are assumed. Also, note that the case $\alpha = 1$ is simply the undiscounted cost problem. J_d is modified into the following form:

$$J_d = E\left[\alpha^n x^T(N)Fx(N) + \sum_{k=0}^{N-1} \alpha^k L(x(k),\ u(k),\ \xi(k))\right] \tag{63}$$

for $0 < \alpha < 1$.

Theorem 10

Consider the discrete-time linear stochastic system given by Eq. (19) for $k = 0, 1, \ldots, N - 1$ and $x(0)$ given, with cost functional given by Eq. (63). Then, the optimal feedback control at each instant of time is given by

$$u(k) = -G(k)x(k), \tag{64}$$

where

$$G(k) = R + \alpha\ E[\Psi^T K(k + 1)\Psi]^{-1} \alpha\ E[\Psi^T K(k + 1)\Phi \tag{65}$$

and

$$\begin{aligned} K(k) = Q &+ \alpha\ E[\Phi^T K(k + 1)\Phi] - \alpha^2\ E[\Phi^T K(k + 1)\Psi] \\ &\times [R + \alpha\ E[\Psi^T K(k + 1)\Psi]]^{-1}\ E[\Psi^T K(k + 1)\Phi] \end{aligned} \tag{66}$$

with

$$K(N) = Q. \tag{67}$$

Now, the optimal average cost is given by the following discounted functional expression:

$$J_d = x^T(0)K(0)X(0) + \sum_{k=0}^{N-1} \mathrm{tr}(\alpha^{k+1} K(k + 1)\Sigma^{\xi\xi}(k)). \tag{68}$$

For a proof see [7].

Remark 5

When $\sqrt{\alpha}\Phi = \Phi'$ and $R/\alpha = R'$ are substituted in Eq. (66), it yields

$$K(k) = Q + E[(\Phi')^T K(k+1)\Phi'] - E(\Phi')^T K(k+1)\Psi] \times [R' + E[\Psi^T K(k+1)\Psi]]^{-1} E[(\Phi')^T K(k+1)\Psi], \quad (69)$$

which has exactly the same form as Eq. (23). Hence, the use of optimal control law (64) [where G(k) is now the constant gain matrix received when the steady-state value of K(k) is substituted for the value of the present equation] will cause the following state equation to evolve:

$$x(k+1) = (\Phi(k) - \alpha\Psi(k)\ E[\Phi^T(k)K(k+1)\Psi(k)] \times [R + \alpha\ E[\Psi^T(k)K(k+1)\Psi(k)]]^{-1})x(k). \quad (70)$$

G. *OPTIMAL CONTROL WITH CORRELATION OF MULTIPLICATIVE AND ADDITIVE NOISES*

Thus far, we have intentionally excluded the possibility of any correlation between the additive system noise vector $\xi(k)$ and the noise components of the stochastic matrices in $\Phi(k)$ and $\Psi(k)$, namely, $\theta(k)$ and $\Gamma(k)$. We will now extend the purely random (white) stochastic problem to include the situation when all the random parameters [except x(0)] are correlated with each other. Furthermore, we assume that the additive noise vector $\xi(k)$ has a nonzero mean given by

$$E[\xi(k)] = \bar{\xi}(k). \quad (71)$$

Let us define the cross covariances between ξ and the stochastic matrices θ and Γ by

$$E\left[\left(\theta_i - \bar{\theta}_i\right)(\xi - \bar{\xi})^T\right] = \Sigma^{\theta_i\xi}, \quad (72)$$

$$E\left[\left(\Gamma_i - \bar{\Gamma}_i\right)(\xi - \bar{\xi})^T\right] = \Sigma^{\Gamma_i\xi}, \quad (73)$$

where θ_i and Γ_i represent the ith columns of θ and Γ, respectively

It has been shown [38] that the covariance matrix of x(k + 1) evolves in the following manner:

$$\Sigma_{ij}^{x(k+1)} = x^T(k)\Sigma^{\Phi_i\Phi_j}x(k) + 2x^T(k)\Sigma^{\Phi_i\Psi_j}u(k)$$
$$+ 2x^T(k)\Sigma^{\Phi_i\xi} + 2u^T(k)\Sigma^{\Psi_i\xi}$$
$$+ u^T(k)\Sigma^{\Psi_i\Psi_j}u(k) + \Sigma^{\xi\xi}. \tag{74}$$

Theorem 11

The optimal control at any instant of time k, for a system governed by the stochastic discrete-time linear equation given in Eqs. (17)-(18) and (71)-(73) is

$$u^*(k) = -G^*(k)x^*(k) - w^*(k), \tag{75}$$

where

$$G^*(k) = [R(k) + \overline{\Psi}^T(k)K(k+1)\overline{\Psi}(k) + \mathrm{Tr}(K(k+1)\Sigma^{\Psi\Psi})]^{-1}$$
$$\times \overline{\Psi}^T(k)K(k+1)\overline{\Phi}(k) + \mathrm{Tr}(K(k+1)\Sigma^{\Phi\Phi}) \tag{76}$$

and

$$w^*(k) = [R(k) + \overline{\Psi}^T(k)K(k+1)\overline{\Psi}(k) + \mathrm{Tr}(K(k+1)\Sigma^{\Psi\Psi})]^{-1}$$
$$\times \overline{\Psi}^T(k)K(k+1)\overline{\xi}(k) + \overline{\Psi}^T(\xi)q(k+1)$$
$$+ \mathrm{Tr}(K(k+1)\Sigma^{\phi\xi}), \tag{77}$$

and where

$$K(k) = Q(k) + \overline{\Phi}^T(k)D(k+1)\overline{\Phi}(k) + \mathrm{Tr}(K(k+1)\Sigma^{\Phi\Phi})$$
$$- [\overline{\Phi}^T(k)K(k+1)\overline{\Psi}(k) + \mathrm{Tr}(K(k+1)\Sigma^{\Phi\Psi})]$$
$$\times [R(k) + \overline{\Psi}^T(k)D(k+1)\overline{\Psi}(k) + \mathrm{Tr}(K(k+1)\Sigma^{\Psi\Psi})]^{-1}$$
$$\times [\overline{\Psi}^T(k)K(k+1)\overline{\Phi}(k) + \mathrm{Tr}(K(k+1)\Sigma^{\Phi\Psi})] \tag{78}$$

and

$$
\begin{aligned}
q(k) = \overline{\Phi}^T(k) &- [\overline{\Phi}^T(k)K(k+1)\overline{\Psi}(k) + \mathrm{Tr}(K(k+1)\Sigma^{\Phi\Psi})] \\
&\times [R(k) + \overline{\Psi}^T(k)K(k+1)\overline{\Psi}(k) + \mathrm{Tr}(K(k+1)\Sigma^{\Psi\Psi})]^{-1}\overline{\Psi}^T \\
&\times [K(k+1)\overline{\xi}(k) + q(k-1)] + \mathrm{Tr}(K(k+1)\Sigma^{\Phi\xi}) \\
&- [\overline{\Phi}^T(k)K(k+1)\overline{\Psi}(k) + \mathrm{Tr}(K(k+1)\Sigma^{\Phi\Psi})] \\
&\times [R(k) + \overline{\Psi}^T(k)K(k+1)\overline{\Psi}(k) \\
&+ \mathrm{Tr}(K(k+1)\Sigma^{\Psi\Psi})]^{-1}\ \mathrm{Tr}(K(k+1)\Sigma^{\Psi\xi}), \qquad (79)
\end{aligned}
$$

with

$$K(N) = F \quad \text{and} \quad q(N) = 0. \qquad (80)$$

As seen from the above equations, the optimal control law now has an additional fixed component with the feedback control. However, the expression for the control gain matrix $G(k)$ is still identical with the one given previously. The implication of the above fact is that the regulation of the state of a linear stochastic system by means of feedback is completely independent of any correlation between the additive and multiplicative noise elements. Furthermore, the stochastic optimal feedback control law is still linear in the state variable.

On the other hand, the correction term $q(k)$ is a function of the cross covariances of the additive and multiplicative noise factors. When the latter covariances are zero, the term $q(k)$ vanishes, and the result reduces to the original problem.

III. UNCERTAINTY MANAGEMENT IN MODELING OF FLEXIBLE STRUCTURES

A. INTRODUCTION

Uncertainty in modeling and control of large flexible structures may arise either from randomness in the properties of the structure itself or from modeling approximations and process idealization. When possible, tests are performed to measure and verify the analytical model of a structure. Model characteristics, that is, frequencies and mode shapes, as well as damping parameters, are compared with the corresponding analytical values and the model is adjusted appropriately. Modeling by finite-element procedures is utilized extensively in the design and development of structures.

Rigid body dynamics are vastly simpler than flexible dynamics, because in the first case only six degrees of freedom are needed to describe the system. Furthermore, the controller logic is relatively simple, since only an inertial model is required for the vehicle. Consequently, the engineer can treat the controllers independently and in a single-input, single-output manner. However, in the case of flexible structures, the number of actuators and sensors must be increased in such a manner that force-motion relationships must reflect elastic and rigid body dynamics. For the above reason, the accuracy of such models has to be treated more carefully.

Various authors have devised methods of generating models for flexible structures that reflect uncertainties [39,40]. However, no realistic approach that addresses modeling of flexible structures by incorporating all errors and randomness within the analytical model has yet been developed.

The present section presents a procedure whereby all uncertainties are incorporated in the best analytical model that is available, and then a stochastic model is generated which reflects uncertainties in the frequencies, mode shapes, and damping ratios.

B. PROBLEM DEFINITION

Differential equations of motion for flexible structures can be represented in the following manner:

$$M\ddot{q} + G\dot{q} + Kq = F, \tag{81}$$

where M is an n × n symmetric positive-definite matrix called the "mass" matrix, G the n × n positive-semidefinite "damping" matrix, K the n × n symmetric positive-semidefinite "stiffness" matrix, q the generalized coordinate vector, and F a forcing function.

Let Φ be the "modal" matrix of the system in Eq. (81); then

$$\Phi^T M \Phi = I, \qquad \Phi^T G \Phi = \text{diag}[-2\zeta_k \delta_k],$$

and

$$\Phi^T K \Phi = \text{diag}\left[\omega_k^2\right], \qquad k = 1, 2, \ldots, n.$$

Let $q = \Phi\eta$; then Eq. (81) can be written as

$$\ddot{\eta} + D\ddot{\eta} + [\omega^2]\eta = \Phi^T F, \tag{82}$$

where $D = \Phi^T G \Phi$. Now, let $x = \begin{pmatrix} \dot{\eta} \\ \eta \end{pmatrix}$; then

$$\dot{x} = Ax + Bu, \tag{83}$$

where

$$A = \begin{bmatrix} [0] & I \\ -[\omega^2] & -[2\zeta\omega] \end{bmatrix}, \qquad B = \begin{bmatrix} [0] \\ \Phi^T F \end{bmatrix}.$$

Obviously, there are three sets of uncertainties one has to consider, namely, ω, ζ, and Φ. Thus, probability density functions (PDFs) of each element of these random matrices should be known a priori. A tremendous amount of data is required to have reasonable PDFs for the above-mentioned parameters. However, there are ways to derive some of them, given partial knowledge. Further research is required in order to reduce the data requirement and facilitate the means of generating the PDFs.

A limited amount of data of frequencies and damping ratios, as well as some mode shapes, were analyzed and possible PDFs were determined for each set. It is well known that frequencies are positive quantities and that lower frequencies are better known than higher ones. The frequencies generated via finite-element methods are (probably) too high for a given mass and stiffness because finite-element models are normally too stiff [41,29]. Moreover, it was found that the PDF that seems to best fit available data of frequencies is the log-normal distribution. In a similar manner, the PDF that seems to fit the data on damping ratios available is the beta distribution.

However, no such conclusion is yet possible for the mode shape uncertainty. The reasons for such a situation are that, first, mode shapes are far more complex than either damping ratios or frequencies. Second, analytical modes (i.e., eigenvectors) are not unique for a given model. Finally, the available data for modes are almost nil and do not support any simplistic conclusions, as was previously done.

Especially important and physically meaningful are the fundamental modes of a structure. Naturally then, classification of structural uncertainties with respect to analysis and specific classes of structures makes a good deal of engineering sense. It is fundamental, for instance, to separate analyses performed for structural integrity from those for jitter or vibration suppression. Similar characteristics probably separate structures of different categories.

C. GENERAL MATRIX THEORY

Any symmetric matrix can be transformed to its Frobenius or companion form through appropriate algorithms [42,30]. For the purpose of generating the PDFs for the elements of the "modal" matrix given the PDFs of the frequencies, the following matrix theory is presented.

Consider the matrix A in companion form

$$A = \begin{bmatrix} 0 & 1 & 0 & 0 & \cdots & & 0 \\ 0 & 0 & 1 & 0 & & & 0 \\ \vdots & & & & \ddots & & \vdots \\ & & & & & & 0 \\ 0 & & \cdots & & & 0 & 1 \\ -\alpha_n & -\alpha_{n-1} & & & \cdots & -\alpha & 1 \end{bmatrix}.$$

Then the characteristic equation for A is

$$\lambda^n + \alpha_1\lambda^{n-1} + \alpha_2\lambda^{N-2} + \cdots + \alpha_n = 0,$$

and if λ_1 is an eigenvalue of A with multiplicity k, then

$$\begin{bmatrix} 1 \\ \lambda_1 \\ \lambda_1^2 \\ \\ \vdots \\ \\ \lambda_1^{n-1} \end{bmatrix} \begin{bmatrix} 0 \\ 1 \\ 2\lambda_1 \\ 3\lambda_1^2 \\ \vdots \\ \\ (n-1)\lambda_1^{n-2} \end{bmatrix} \begin{bmatrix} 0 \\ 0 \\ 1 \\ \\ \vdots \\ \\ (n_2-1)\lambda_1^{n-3} \end{bmatrix} \begin{bmatrix} 0 \\ 0 \\ 0 \\ 1 \\ \vdots \\ \\ (n_3-1)\lambda_1^{n-4} \end{bmatrix} \cdots \begin{bmatrix} 0 \\ 0 \\ 0 \\ \\ \vdots \\ 0 \\ (n_k-1)\lambda_1^{n-k+1} \end{bmatrix} \tag{84}$$

are generalized eigenvectors of A associated with λ_1 where

$$\binom{n-1}{i} = \frac{(n-1)(n-2)\cdots(n-i)}{1 \times 2 \times 3 \times \cdots \times i}.$$

Thus, in case λ_i has multiplicity 1 for all i = 1, ..., n, the "modal" matrix of A can be written as follows:

$$\Phi = \begin{bmatrix} 1 & 1 & \cdots & 1 \\ \lambda_1 & \lambda_2 & & \lambda_n \\ \lambda_1^2 & \lambda_2^2 & & \lambda_n^2 \\ \vdots & & & \vdots \\ \lambda_1^{n-1} & \lambda_2^{n-1} & & \lambda_n^{n-1} \end{bmatrix} \tag{85}$$

which is the *Vandermonde* matrix.

The above matrix can be a very ill-conditioned matrix, especially for relatively large values of λ_i's. However, for small-order systems it might be a utilizable approach. Thus, statistically one can generate the PDFs of all the elements of Φ if the PDFs of the ω's are known. All the statistical information regarding uncertainties in the model of a structure can be incorporated in a stochastic model, thus yielding a more realistic representation of the structure, on the average.

From Eqs. (22) and (23) it is apparent that the following statistical expressions have to be evaluated in order to compute the optimal control: $E[\Psi^T K \Psi]$, $E[\Psi^T K \Phi]$, $E[\Phi^T K \Phi]$. It will entail some complex algebra to compute the above expressions given the required PDFs. However, for nominally small systems it can be achieved without major difficulty.

IV. OPTIMAL ESTIMATION OF THE STATES OF LINEAR DISCRETE STOCHASTIC SYSTEMS

In situations in which the entire state vector cannot be measured, as is typical in almost all complex systems, the control law deduced in the form of u(k) = f(x(k), k) cannot be implemented. Thus, either we have to directly account for the unavailability of the entire state vector or an appropriate approximation to the state vector must be determined that can be substituted into the control law. In almost all cases it is simpler to use an approximation to the state vector than a new direct attack on the design problem. The above-mentioned replacement of the unavailable state vector by an approximation results in the decomposition of a control design problem into two separate phases. The first phase is designing the control law with the assumption that the state vector is available (which usually results in a control law without dynamics). The second phase is the design of a system that produces an approximation to the state vector. This new system, which is called an observer, has as its inputs the inputs and available output of the system whose state is to be approximated and has a state vector that is linearly related to the desired approximation. A detailed study of observers is found in D. Luenberger's paper [43].

For the standard LQG problem, the optimal stochastic control problem is separable into the optimal deterministic control and optimal estimation without control. The duality theorem [44] shows how one problem is the dual of the other. The truly optimal filter for the problem at hand will be nonlinear and infinite dimensional due to the product of random vectors and random matrices. However, the optimal estimation algorithm that is presented in this section is only optimal in the class of *linear* estimators. The necessary assumptions required to derive the *linear unbiased* estimator automatically renders the filtering algorithm only *suboptimal*.

In this section we will use the idea of direct derivation of the optimal linear filter developed in Athans and Tse's paper [45] and will construct an algorithm which is of lower dimension than the Kalman-Bucy filtering algorithm. Petrov and Minin [46] in their paper developed a similar algorithm for the case of linear stochastic systems with only additive noise involved. Our effort can be regarded as a generalization of the theory of observers for our problem.

A. PROBLEM STATEMENT

The controlled linear stochastic dynamical system is described by Eq. (17). Here, the initial state $x(0)$ is a random Gaussian vector independent of all the elements of the random matrices and vectors with

$$E[x(0)] = \bar{x}_0 \quad \text{and} \quad E\left[\left(x(0) - \bar{x}_0\right)\left(x(0) - \bar{x}_0\right)^T\right] = \bar{X}_0. \tag{86}$$

In the sequel we assume that

$$E[x(0)\xi^T(k)] = 0 \quad \text{for} \quad k = 0, 1, \ldots, N. \tag{87}$$

Similarly, $\xi(k)$ is independent of all the elements of $\theta(k)$ and $\Gamma(k)$ for all $k = 0, 1, \ldots, N$. The system measurements are given by

$$y(k) = \begin{bmatrix} y_1(k) \\ y_2(k) \end{bmatrix} = \begin{bmatrix} C(k) \\ \Omega(k) \end{bmatrix}\begin{bmatrix} X_1(k) \\ X_2(k) \end{bmatrix} + \begin{bmatrix} 0 \\ \bar{\nu}(k) \end{bmatrix}, \tag{88}$$

where $y_1(k)$ is the l-dimensional vector of exact measurements, $y_2(k)$ is the q-dimensional vector of noisy measurements, $C(k)$ is the $l \times n$ deterministic matrix, and $\Omega(k)$ the $q \times n$ stochastic matrix of Gaussian elements having given statistics as follows:

$$\begin{aligned} E[\Omega(k)] &= \bar{\Omega}(k) \quad \text{and} \\ E[(\Omega(k) - \bar{\Omega}(k))(\Omega(r) - \bar{\Omega}(r))^T] &= \Sigma^{\Omega\Omega}(k)\delta_{kr}. \end{aligned} \tag{89}$$

Finally, $\nu(k)$ is a vector of Gaussian white noise with dimensions of $q \times l$ and statistics of

$$E[\nu(k)] = 0 \quad \text{and} \quad E[\nu(k)\nu^T(r)] = \Sigma^{\nu\nu}(k)\delta_{kr} \tag{90}$$

and

$$E[\nu(k)\xi^T(r)] = \Sigma^{\nu\xi}(k)\delta_{kr}. \tag{91}$$

A problem of the above kind occurs in practice when, as mentioned earlier, some measurements of a system are taken with non-white noise and some with white noise.

The optimal control problem is to find a control $u(k)$, $k = 0, 1, \ldots, N$, which will minimize the performance criterion in Eq. (5).

B. *REFORMULATION OF THE PROBLEM*

The solution to the stochastic optimal control problem with multiplicative and additive random parameters shows the effects of uncertainties in the performance of the control system. Here the control gain will be a function of the unconditional means and covariances of the uncertainties. Hence, the suboptimal stochastic control law is considered based upon the open-loop feedback method [47]. Thus a control of the form

$$u(k) = -G(k)x(k) \tag{92}$$

is assumed and $x(k)$ is found by means of mathematical Kalman-Bucy-type filtering algorithms [25].

In the following pages we will construct a filtering algorithm similar to, but of lower dimension than, the Kalman-Bucy filtering algorithm. While the Kalman-Bucy filter for our problem is n dimensional, the algorithm we will develop will perform the filtering with a $(n - l)$-dimensional filter. A similar lowering of dimension is possible for the noiseless measurement vector $y_1(k)$ given in Eq. (88) [46]. A system will be constructed in which the optimal mean-square estimate $\hat{x}(k)$ of the state vector $x(k)$ will be derived, the computational as well as storage requirements of which will be less than that of the Kalman-Bucy filtering algorithm. We will call the above-mentioned system "the observer," following Luenberger [43].

Let us select a vector $z(k)$ of dimension $(n - l)$ such that the augmented vector

$$\begin{bmatrix} z(k) \\ y_1(k) \end{bmatrix}$$

will be connected with $x(k)$ by a nondegenerate transformation $M(k)$ by the following equation:

$$\begin{bmatrix} z(k) \\ y_1(k) \end{bmatrix} = M(k)x(k), \tag{93}$$

where

$$\det M(k) \neq 0 \quad \text{and} \quad M(k) = \begin{bmatrix} M_1(k) \\ C(k) \end{bmatrix} \tag{94}$$

and where $M_1(k)$ is an $(n - l) \times n$-dimensional matrix and det stands for the determinant operator.

Hence, from Eq. (93) we have

$$x(k) = M^{-1}(k)\begin{bmatrix} z(k) \\ y_1(k) \end{bmatrix}. \tag{95}$$

Substituting Eq. (95) into Eq. (17) yields

$$\begin{aligned} x(k+1) &= M^{-1}(k+1)\begin{bmatrix} z(k+1) \\ y_1(k+1) \end{bmatrix} \\ &= \begin{bmatrix} A(k) & 0 \\ 0 & \theta(k) \end{bmatrix}\begin{bmatrix} M_1(k) \\ C(k) \end{bmatrix}^{-1}\begin{bmatrix} z(k) \\ y_1(k) \end{bmatrix} + \begin{bmatrix} B(k) \\ \Gamma(k) \end{bmatrix}u(k) + \begin{bmatrix} 0 \\ \xi(k) \end{bmatrix}. \end{aligned} \tag{96}$$

Rewriting Eq. (96), we have

$$\begin{bmatrix} z(k+1) \\ y_1(k+1) \end{bmatrix} = M(k+1)\begin{bmatrix} A(k) & 0 \\ 0 & \theta(k) \end{bmatrix}\begin{bmatrix} M_1(k) \\ C(k) \end{bmatrix}^{-1}\begin{bmatrix} z(k) \\ y_1(k) \end{bmatrix}$$

$$+ M(k+1)\begin{bmatrix} B(k) \\ \Gamma(k) \end{bmatrix}u(k) + M(k+1)\begin{bmatrix} 0 \\ \xi(k) \end{bmatrix}. \quad (97)$$

We can now write Eq. (97) in a more compact form as

$$z(k+1) = A_{11}(k, k+1)z(k) + A_{12}(k, k+1)y_1(k)$$
$$+ B_1(k, k+1)u(k) + \xi_1(k), \quad (98a)$$

$$y_1(k+1) = A_{21}(k, k+1)z(k) + A_{22}(k, k+1)y_1(k)$$
$$+ B_2(k, k+1)u(k) + \xi_2(k), \quad (98b)$$

where

$$\begin{bmatrix} A_{11} & A_{12} \\ A_{21} & A_{22} \end{bmatrix} = \begin{bmatrix} M_1(k+1) \\ C(k+1) \end{bmatrix}\begin{bmatrix} A(k) & 0 \\ 0 & \theta(k) \end{bmatrix}\begin{bmatrix} M_1(k) \\ C(k) \end{bmatrix}^{-1}, \quad (99)$$

$$\begin{bmatrix} B_1(k, k+1) \\ B_2(k, k+1) \end{bmatrix} = M(k+1)\begin{bmatrix} B(k) \\ \Gamma(k) \end{bmatrix} \quad \text{and}$$

$$\begin{bmatrix} \xi_1(k) \\ \xi_2(k) \end{bmatrix} = M(k+1)\begin{bmatrix} 0 \\ \xi(k) \end{bmatrix}. \quad (100)$$

In the above equations, A_{11}, A_{12}, A_{21}, A_{22}, B_1, and B_2 are $(n - l) \times (n - l)$, $(n - l) \times l$, $l \times (n - l)$, $l \times l$, $(n - l) \times lm$, and $l \times m$ matrices, respectively, and $\xi_1(k)$, $\xi_2(k)$ are $(n - l)$- and l-dimensional vectors of Gaussian white noises whose statistical properties can be derived from the appropriate relations.

Consider Eq. (98b). We can rewrite it as

$$z_2(k) \triangleq \Omega(k)x(k) + x(k) = \Omega(k)M^{-1}(k)\begin{bmatrix} z(k) \\ y_1(k) \end{bmatrix} + \nu(k), \quad (101)$$

and representing $\Omega(k)M^{-1}(k)$ by the system

$$\Omega(k)M^{-1}(k) = [H_1(k) : H_2(k)], \tag{102}$$

where $H_1(k)$ and $H_2(k)$ are $q \times (n - l)$- and $q \times l$-dimensional matrices, respectively, we obtain, after rearranging terms,

$$z_3(k) \triangleq y_2(k) - H_2(k)y_1(k) = H_1(k)z(k) + \nu(k). \tag{103}$$

Since $y_2(k)$ and $y_1(k)$ are available through Eqs. (101) and (103), we can assume that there is a vector of measurements given by

$$z_3(k) = H_1(k)z(k) + \nu(k). \tag{104}$$

The new problem reduces to estimating the vector $z(k)$ given by the first equation in the system [Eq. (98b)] with the measurements given by Eqs. (101) and (104).

To simplify matters further, we will introduce the following $(l + q)$-dimensional augmented measurements:

$$z_a(k) \triangleq \begin{bmatrix} z_2(k) \\ z_3(k) \end{bmatrix} = H_a(k, k+1)z(k) + \xi_a(k), \tag{105}$$

where

$$H_a(k, k+1) = \begin{bmatrix} A_{22}(k, k+1) \\ H_1(k, k+1) \end{bmatrix} \quad \text{and}$$

$$\xi_a(k) = \begin{bmatrix} \xi_2(k) \\ \nu(k) \end{bmatrix}. \tag{106}$$

C. EQUATION OF THE OBSERVER

For the newly developed system given by

$$z(k+1) = A_{11}(k, k+1)z(k) + A_{12}(k, k+1)y_1(k)$$
$$+ B_1(k, k+1)u(k) + \xi_1(k), \tag{107}$$

$$z_a(k) = H_a(k, k+1)z(k) + \xi_a(k).$$

we will write the equation of the observer following the procedures in [45] and [46]:

$$\hat{z}(k+1) \triangleq K_1(k)\hat{z}(k) + K_2(k)z_a(k) + K_3(k)[B_1(k, k+1)u(k) + A_{12}(k, k+1)y_1(k)]. \tag{108}$$

In Eq. (108), K_1, K_2, and K_3 are $(n - l) \times (n - l)$-, $(n - l) \times (l + q)$-, and $(n - l) \times (n - l)$-dimensional matrices of unknown observer coefficients, respectively.

Define the error in estimation by

$$e(k) = z(k) - \hat{z}(k). \tag{109}$$

By direct application of Eqs. (107) and (108), Eq. (109) takes the following form:

$$\begin{aligned} e(k+1) = {} & [A_{11}(k, k+1) - K_1(k) - K_2(k)H_a(k, k+1)]z(k) \\ & + K_1(k)e(k) - K_2(k)\xi_a(k) + [I_{n-1} - K_3(k)] \\ & \times [B_1(k, k+1)u(k) + A_{12}(k, k+1)y_1(k)] \\ & + \xi_1(k), \end{aligned} \tag{110}$$

where I_{n-l} represents the $(n - l) \times (n - l)$-dimensional identity matrix.

Considering the fact that we want to develop an unbiased linear estimator, we should guarantee the following:

$$E[e(0)] = 0 \quad \text{and} \quad E(e(k+1)] = 0 \quad \text{for} \quad k = 0, 1, \ldots, N. \tag{111}$$

Through Eq. (111) we obtain

$$K_3(k) = I_{n-l} \quad \text{and} \quad K_1(k) = E[A_{11}(k, k+1)] - K_2(k)\, E[H_a(k, k+1)], \tag{112}$$

and, consequently, the observer equation (108) reduces to

$$\hat{z}(k + 1) = (E[A_{11}(k, k + 1)] - K_2(k)\ E[H_a(k, k + 1)])\hat{z}(k)$$
$$+ K_2(k)z_a(k) + B_1(k, k + 1)u(k)$$
$$+ A_{12}(k, k + 1)y_1(k), \qquad (113)$$

with its initial condition represented by

$$\hat{z}(0) = M_1(0)\ E[x(0)] = M_1(0)\overline{x}_0. \qquad (114)$$

The error in estimation at the (k + 1)th interval can be expressed through Eqs. (107) and (113) as

$$z(k + 1) - \hat{z}(k + 1) = A_{11}(k, k + 1)z(k) - E[A_{11}(k, k + 1)]$$
$$- K_2\ E[H_a(k, k + 1)]\hat{z}(k)$$
$$+ K_2 a_a(k) + \xi_1(k)$$
$$= A_{11}(k, k + 1)e(k) - (E[A_{11}(k, k + 1)]$$
$$- A_{11}(k, k + 1) - K_2\ E[H_a(k, k + 1)])\hat{z}(k)$$
$$- K_2 H_a(k, k + 1)z(k)$$
$$- K_2\xi_a(k) + \xi_1(k). \qquad (115)$$

The error covariance matrix of our system can now be defined in the following manner:

$$P(k + 1) = E[e(k + 1)e^T(k + 1)]. \qquad (116)$$

The problem of obtaining the optimal mean-square estimate of x(k) requires, as a necessary condition, the minimization of the functional tr(P(k + 1)) with respect to K_2 [46].

Then, solving for K_2, after some mathematical manipulations, we obtain

$$K_2(k) = \frac{1}{2}\left[A_{11}\ E[zz^T]\bar{A}_{11}^T + Tr(\Sigma^{A_{11}A_{11}}\ E[zz^T]) + 2\bar{A}_{11}\ E[zz^T]\bar{H}_a^T + 2\Sigma^{\xi_1\xi_a}\right]\left[\bar{H}_a^T\ E[zz^T]^T\bar{H}_a + \Sigma^{\xi_a\xi_a}\right]^{-1}. \quad (117)$$

The coefficient matrix given by Eq. (117) minimizes the trace of the error covariance matrix. In other words, it minimizes the dispersion of the error at the next step. The initial condition for the error covariance matrix is determined via the initial conditions on z(0) in Eq. (114). Thus, the initial condition for Eq. (116) is

$$P(0) = E\left[\left[z(0) - M_1(0)\bar{x}_0\right]\left[z(0) - M_1(0)\bar{x}_0\right]^T\right]. \quad (118)$$

The optimal observer equation is determined by Eqs. (113), (114), (116), (117), and (118).

According to Eq. (95) the optimal mean-square estimate of the state variable will be given by

$$x(k) = M^{-1}(k)\begin{bmatrix} z(k) \\ y_1(k) \end{bmatrix}. \quad (119)$$

In the above, the need to know that $y_1(k + 1)$ at time $k + 1$ for estimating the kth estimate is not desirable. However, it is possible to generate real-time estimates by performing a minor conversion. For details of the above conversion see [8].

V. OPTIMAL CLOSED-LOOP CONTROL OF STOCHASTIC SYSTEMS WITH NOISY MEASUREMENTS

In the usual deterministic linear quadratic control problem, it is well known [18] that a unique solution exists if the system is asymptotically stable. In linear stochastic discrete-time systems with random parameters, the first and second moments can be studied for stabilizability, detectability, and observability criteria [10].

A. PROBLEM STATEMENT

Consider a linear discrete system with random parameters given by

$$x_{i+1} = \Phi_i x_i + \Gamma_i u_i, \tag{120}$$

where $x_i \in R^n$ is the state, $u_i \in R^m$ the control, and $\Phi_i \in M^{n\times n}$, $\Gamma_i \in M^{n\times m}$, where M is the Banach space of all real matrices with the usual norm. In the above $\{\Phi_i\}$ and $\{\Phi_i\}$ are sequences of independent random matrices with stationary statistics.

The observability criteria is just an extension of the observability criteria for deterministic systems essentially with the matrices replaced by their mean values (for details see [10]).

If the system in Eq. (120) is mean-square stabilizable, then the infinite-time discrete optimal control problem has a solution. Moreover, if $(\Phi, Q^{1/2})$ is mean-square observable, where Q is the weighting matrix given in Eq. (15), then the solution for the optimal control problem is unique.

Detectability conditions can similarly be derived. Thus, the system (120) with measurements given in Eq. (88) is detectable in the mean if

$$E[y_i] = 0 \quad \forall_i \quad \Rightarrow \quad E[x_i] \to 0$$

and is detectable in the mean-square sense if

$$E\left[\|y_i\|^2\right] = 0 \quad \forall_i \quad \Rightarrow \quad E\left[\|x_i\|^2\right] \to 0.$$

Moreover, observability of the system (120) with its measurements in the mean implies mean detectability. Similarly, mean-square observability implies mean-square detectability [48].

It is also possible to show [49] that the optimal linear time-invariant estimate for a system like Eq. (120) exists, is stable, and is unique if the system is stable in the mean-square sense.

B. *OPTIMAL LINEAR STOCHASTIC DISCRETE-TIME PROBLEMS WITH INCOMPLETE STATE INFORMATION*

Let us define the observed outputs by

$$y_k = [y^T(0),\ y^T(1),\ \ldots,\ y^T(k)]^T. \tag{121}$$

Moreover, assume that the admissible controls are functions of outputs y_{k-1}.

It can be shown [7] that the minimum value of the cost functional given in Eq. (5) is

$$\begin{aligned} J^* = E\Big[& x^T(0)K(0)x(0) + \operatorname{tr}\big(\overline{x}_0 K(0)\big) \\ & - \sum_{k=0}^{n-1} \{\operatorname{tr} \Sigma^{xx} G^T [R + \overline{\Psi}^T K(k+1)\overline{\Psi} + \operatorname{tr}(\Sigma^{\Psi\Psi} K(k+1))]G \\ & - \operatorname{tr}(\Sigma^{\xi\xi} K(k+1)) + x^T \overline{\Phi}^T K(k+1)\overline{\Psi}^T K(k+1)\overline{\Phi} x \\ & + 2x^T \operatorname{Tr}(\Sigma^{\Phi\Psi} K(k+1)) G x\} \,|\, y_{k-1}\Big], \end{aligned} \tag{122}$$

where

$$x(k) = E[x(k)\,|\,y_{k-1}], \tag{123}$$

$$\Sigma^{xx}(k) = E[(x(k) - x(k))(x(k) - x(k))^T],$$

and $G(k)$ is as given previously.

VI. CONCLUSIONS

Optimization and control of stochastic dynamic systems under uncertainty possess several important characteristics that are absent where complete certainty prevails. During the decision process regarding an optimal policy, the possibility of feedback (information) and the risk in proper modeling and formulation of the problem must both be taken into account. The stochastic formulation with stochastic time-dependent or constant parameters allows the simultaneous treatment of several important classes of problems in various fields, such as stochastic adaptive control and control of finite-state Markov chains as well as aerospace and macroeconomic control.

The main objective of our article has been to provide a brief overview of a very important problem in multidimensional stochastic control problems. Moreover, a unified framework for optimization and control of linear discrete-time multidimensional stochastic systems under partial information has been treated in a rather concise manner. It is shown that the control gains under the above-mentioned situation depend simultaneously upon the mean values and covariances of the random variables or parameters involved. Thus, the effect of parameter/random disturbances is manifested through modulation of the control gains of the system.

For a stochastic control system, it is required that the controller be able to reduce the effect of uncertainty in the states and regulate the process by adapting to the random variations taking place. Moreover, the controller must perform the above-mentioned action without any learning, thus being only adaptive in the narrow sense of the word. Hence, both the stochastic optimal control and estimation have to be dealt with simultaneously. However, under partial information, it is essential to balance identification/estimation and control, thus providing the appropriate suboptimal solution with the given performance criterion. One way of accomplishing the above balancing is by fixing the structure of the controller, as is done in [7] and [8]. Thus, taking a linear controller with a linear estimator, joint optimization of the control and filtering gains of an inherently nonlinear problem is accomplished through the optimization of the quadratic cost functional.

A different notion of strong stabilizability referred to as "structural stabilizability via a nominally determined quadratic Lyapunov function" is presented in [50]. Moreover, in [50] the uncertainties are bounded and the stability conditions are treated under direct feedback control.

REFERENCES

1. B. WITTENMARK, "Stochastic Adaptive Control Methods: A Survey," *Int. J. Control 21*, No. 5, 705-730 (1975).

2. R. BELLMAN, "Adaptive Control Processes: A Guided Tour," Princeton Univ. Press, Princeton, New Jersey, 1961.

3. M. AOKI, "Optimization of Stochastic Systems," Academic Press, New York, 1967.

4. M. ATHANS, "The Role and Use of the Stochastic LQG Problem in Control System Design," *IEEE Trans. Autom. Control AC-16*, No. 6, 529-552, December 1971.

5. O. L. MANGASARIAN, "Nonlinear Programming," McGraw-Hill, New York, 1969.

6. J. STERNBY, "A Simple Dual Control Problem with an Analytical Solution," *IEEE Trans. Autom. Control AC-21*, 840-844, December 1976.

7. H. PANOSSIAN, "Optimal Adaptive Stochastic Control of Linear Dynamical Systems with Multiplicative and Additive Noise," Ph.D. Dissertation, University of Calfiornia, Los Angeles, 1981.

8. H. V. PANOSSIAN and C. T. LEONDES, "Observers for Optimal Estimation of the State of Linear Stochastic Discrete Systems," *Int. J. Control 37*, No. 3, 645-655 (1983).

9. J. ZABCZYK, "On Optimal Stochastic Control of Discrete-Time Systems in Hilbert Spaces," *SIAM J. Control 13*, No. 6, November 1975.

10. W. L. DeKONING, "Infinite Horizon Optimal Control for Linear Discrete Time Systems with Stochastic Parameters," *Automatica 18*, No. 4, 443-453 (1982).

11. S. E. HARRIS, "Stochastic Controllability of Linear Discrete Systems with Multiplicative Noise," *Int. J. Control 27*, No. 2, 213-227 (1978).

12. V. I. IVANENKO, "Control in the Case of Stochastic Indeterminacy," *Sov. Autom. Control (Engl. Transl.) 15*, No. 6, 63-71 (1982).

13. A. T. BHARUCHA-REID (ed.), "Probabilistic Methods in Applied Mathematics," Vols. I-III, Academic Press, New York, 1970.

14. M. L. MEHTA, "Random Matrices and Statistical Theory of Energy Levels," Academic Press, New York, 1967.

15. N. WIENER, "Extrapolation, Interpolation and Smoothing of Stationary Time Series," Wiley, New York, 1949.

16. W. M. WONHAM, "Random Differential Equations in Control Theory," *in* "Probabilistic Methods in Applied Mathematics," Vol. 2 (A. T. Bharucha-Reid, ed.), pp. 131-212, Academic Press, New York, 1970.

17. P. J. McLANE, "Linear Optimal Stochastic Control Using Instantaneous Output Feedback," *Int. J. Control 13*, No. 2, 383-396 (1971).

18. H. KWAKERNAAK and R. SIVAN, "Linear Optimal Control Systems," Wiley (Interscience), New York, 1972.

19. K. Y. LEE, S. N. CHOW, and R. D. BARR, "On the Control of Discrete Time Distributed Parameter Systems," *in* "Recent Developments in Control Theory," (N. P. Bhatia, ed.), pp. 141-156, SIAM, 1972.

20. A. E. BRYSON and Y. C. HO, "Applied Optimal Control," Ginn (Blaisdell), Boston, Massachusetts, 1969.

21. M. ATHANS and P. L. FALB, "Optimal Control," McGraw-Hill, New York, 1966.

22. H. PAYNE and L. M. SILVERMAN, "On the Discrete Time Algebraic Riccati Equation," *IEEE Trans. Autom. Control AC-18*, No. 3, 226-234, June 1973.

23. G. A. HEWER, "Analysis of a Discrete Matrix Riccati Equation of Linear Control and Kalman Filtering," *J. Math. Anal. Appl. 42*, 226-236 (1973).

24. W. W. HAGER and L. K. HOROWITZ, "Convergence and Stability Properties of the Discrete Riccati Operator Equation and Associated Optimal Control and Filtering Problems," *SIAM J. Control Optim. 14*, No. 2, February 1976.

25. R. KU, "Adaptive Stochastic Control of Linear Systems with Random Parameters," Ph.D. Dissertation, MIT Press, Cambridge, Massachusetts, May 1978.

26. M. AOKI, "Control of Large-Scale Dynamic Systems by Aggregation," *IEEE Trans. Autom. Control AC-13*, No. 3, June 1968.

27. B. FRIEDLAND, "Treatment of Bias in Recursive Filtering," *IEEE Trans. Autom. Control AC-14*, No. 4, August 1969.

28. A. K. MAHALANABIS and K. RANA, "Guaranteed Cost Solution of Optimal Control and Game Problems for Uncertain Systems," *in* "Optimal Control Applications and Methods," pp. 353-360, 1980.

29. M. AOKI, "Control of Linear Discrete-Time Stochastic Dynamic Systems with Multiplicative Disturbances," *IEEE Autom. Control AC-20*, 388-392, June 1975.

30. T. KATAYAMA, "On the Matrix Riccati Equation for Linear Systems with a Random Gain," *IEEE Trans. Autom. Control AC-21*, 770-771, October 1971.

31. S. NIWA, M. HAYASE, and I. SUGIURA, "Stability of Linear Time-Varying Systems with State Dependent Noise," *IEEE Trans. Autom. Control AC-21*, 775-776, October 1976.

32. J. L. WILLEMS and J. G. WILLEMS, "Feedback Stabilizability for Stochastic Systems with State and Control Dependent Noise," *Automatica 12*, 277-283, May 1976.

33. F. KOZIN, "A Survey of Stability of Stochastic Systems," *Automatica 5*, 95-112, January 1969.

34. V. M. KONSTANTINOV, "The Stability of Stochastic Difference Systems," *Probl. Inf. Transm. (Engl. Transl.) 6*, 70-75, January-March 1970.

35. M. H. A. DAVIS, "Linear, Estimation and Stochastic Control," Chapman & Hall, London, 1977.

36. G. C. CHOW, "Analysis and Control of Dynamic Economic Systems," Wiley, New York, 1975.

37. H. J. KUSHNER, "Probability Methods for Approximations in Stochastic Control for Elliptic Equations," Academic Press, New York, 1977.

38. F. R. SHUPP, "Uncertainty and Optimal Policy Intensity of Fiscal and Income Policies," *Annu. Econ. Soc. Meas. 5/2*, 225-237 (1976).

39. G. C. HART and J. D. COLLINS, "The Treatment of Randomness in Finite Element Modeling," *SAE Shock Vib. Symp.*, Los Angeles, California, October 1970.

40. T. K. HASSELMAN, "Structural Uncertainty in Dynamic Analysis," *SAE Shock Vib. Conf.*, Los Angeles, California, 1982.

41. O. C. ZIENKIEWICZ, "The Finite Element Method," 3rd ed., McGraw-Hill, New York, 1977.

42. A. S. HOUSEHOLDER, "The Theory of Matrices in Numerical Analysis," Dover, New York, 1975

43. D. G. LUENBERGER, "An Introduction to Observers," *IEEE Trans. Autom. Control AC-16*, No. 6, 596-602, December 1971.

44. A. A. FELDBAUM, "Optimal Control Systems," Academic Press, New York, 1965.

45. M. ATHANS and E. TSE, "A Direct Derivation of the Optimal Linear Filter Using the Max Principle," *IEEE Trans. Autom. Control AC-12*, No. 6, 690-694, December 1967.

46. A. I. PETROV and V. V. MININ, "Application of Observers for Optimal Control of Linear Stochastic Discrete Systems," *Sov. Autom. Control (Engl. Transl.)*, 29-35 (1979).

47. R. KU and M. ATHANS, "On the Adaptive Control of Linear Systems Using Open-Loop Feedback Optimal Approach," *IEEE Trans. Autom. Control AC-18*, 489-493, October 1973.

48. W. L. DeKONING, "Detectability of Linear Discrete-Time Systems with Stochastic Parameters," *Int. J. Control 38*, No. 5, 1035-1046 (1983).

49. W. L. DeKONING, "Optimal Estimation of Linear Discrete-Time Systems with Stochastic Parameters," *Automatica 20*, No. 1, 113-115 (1984).

50. I. R. PETERSEN, "Structural Stabilization of Uncertain Systems: Necessity of the Matching Condition," *SIAM J. Control Optim. 23*, No. 2, March 1985.

Multicriteria Optimization in Adaptive and Stochastic Control

NICK T. KOUSSOULAS

Bell Communications Research
331 Newman Springs Road
Red Bank, New Jersey 07701

I. INTRODUCTION

The central problem in the conception of a parameter adaptive or a (suboptimal) stochastic control scheme is the balance between the two conflicting roles of the input, namely, learning and control. The former task may require large-amplitude wide-bandwidth signals, something that can prove detrimental for the latter since the system state may be driven away from the desired one or instability may even occur. An input that exploits the possibility of learning (whenever it exists) is called a control with dual effect, usually referred to simply as "dual control." It is possible, of course, to ignore this interaction and enforce certainty equivalence or control-estimation separation. In the case of linear systems with unknown parameters, the resulting simplicity, combined with the availability of fast, cheap, and reliable digital processors, makes the implementation of such algorithms possible. Two of the most successful schemes along these lines are the self-tuning regulator (STR) and the model-reference adaptive control (MRAC). The STR has known wide applicability, mainly in the process industry, and is the first adaptive control scheme to be mass marketed, with several variations and improvements incorporated on the way to deal (*ad hoc*, mostly) with delays, nonlinearities, load variations, and/or other situations that violate its inherent initial assumptions [1].

On the other hand, adaptive control schemes have been proposed that acquire an element of duality by taking account of the effect of input upon the estimation function. One of the first works in this spirit seems to be [2], where a measure of the estimator's performance one step ahead is included in the cost to be minimized, forcing the control to affect the quality of the estimation process. The idea has been further exploited in the following years [3,4,5]. Recent developments in modern control theory were also used to formulate the (adaptive) dual-control problem. Hierarchical and multilevel control concepts as well as multiobjective optimization (ϵ constraint method) were used in [6] for the simultaneous identification and control of deterministic systems. The same philosophical basis was the foundation of [7] and [8], where an on-line solution was proposed. Another recent contribution is [9], where decoupling is used to optimize a hierarchically formulated performance index, thus providing the controls for a system in steady state (no dynamic models are considered); model mismatching is taken account of, making the method similar to output error identification.

In the area of stochastic control, that is, when the system parameters are known, several schemes have been invented that try to approximate the true dual-control solution of Feld'baum [10], which, for nontrivial systems, can be considered impossible to compute with presently available means. Depending on the use of the available information the solution is called open-loop feedback-optimal, feedback, or closed-loop [11]. Complexity and computational load problems plague these schemes as well.

In this article we present a method of balancing control and learning using a multicriteria optimization concept. The resulting scheme can be incorporated in the structure of virtually all existing adaptive control algorithms in order to improve their learning capabilities. Application to stochastic control algorithms is also possible. The basic idea is to force the closed-loop system performance to move toward a specified set of goals for the controller, in the control part, and the parameter and/or state estimator, in the learning part. An appropriate name for this new approach, then, would be "adaptation by targets." The following sections contain a description of the models, control criteria, and estimation procedures, a detailed presentation of the basic principle, the techniques for target selection, examples, and comments on the performance of the algorithm.

II. MODELS, CRITERIA, AND ESTIMATORS

Before we proceed with the detailed description of the algorithm we shall cursively review the basic steps in the design of an adaptive controller. We can distinguish three "building blocks": system model, estimator, and control criterion. If we limit the discussion to linear time-invariant systems, there are two approaches. First, the input-output (ARMA) formulation [12,13], which results in (pseudo)linearity and, if only output regulation is required, has the advantage of no need for state estimation. A further reduction in the number of parameters can be achieved through a judicious choice of the identification method. Second, the state-space formulation allows more general models to be considered, namely, those where the measurement-contaminating noise is different than the plant noise. Moreover, the state-space structure allows a "natural" approach to the parameter estimation problem through state augmentation. The price to pay for this unification and generality is increased complexity, which leads to difficulty of analysis and computational and implementation problems.

The corresponding identification and state estimation procedures, called from this point on simply estimation, form two different groups, even though the functional principles are quite similar. Recursive parameter estimation methods like least squares, extended least squares, maximum likelihood, or stochastic approximation are more suitable for input-output models. The extended Kalman filter (EKF), possibly a modified one [14], befits state-space models. Nonlinear estimation schemes may, in principle, be used as well, since the problem, even in the linear time-invariant case, is nonlinear.

The last design element is the selection of the control criterion. There are usually two choices: (1) the minimum output variance and (2) the weighted quadratic penalty on the state and control energies, with a horizon of one or more steps. They are perfectly matched with the input-output model and the state-space formulation, respectively.

While the above barely sketch the options for the building of an adaptive control scheme, detailed exposition of the formulation, criteria and procedures used in adaptive control is not within the scope of this article; the interested reader may refer, for example, to [10,15] and if necessary to the references therein. In general, any combination of a model, an estimator, and a control policy related to a criterion can qualify, in principle, as an adaptive controller. Practical considerations, ease of implementation, and above all (apparent) stability and satisfactory performance will dictate the final choice.

III. BASIC PRINCIPLE AND ALGORITHM DESCRIPTION

The natural setting where the multicriteria optimization (MCO) was created and evolved is the satisfaction of conflicting requirements. A rapid development of methodologies can be observed the past few decades [16], accompanied by a significant increase in the number of applications. Even so, the involvement of control engineers and researchers has been largely limited so far to theoretical issues and actual applications are few and far between. The MCO issues pertinent to the control of deterministic dynamic systems are contained in [17]. Extension to stochastic problems involving linear systems only (LQG-type problems) has been provided in [18] and [19].

A multicriteria optimization methodology that has been shown to be applicable to control systems problems is the ideal point method. In this method, the costs that result from the individual optimization of each scalar criterion separately define an ideal or utopian point in the cost space, which is subsequently approximated through the minimization of a "global" criterion of the form

$$R_p = \|J - J^*\|_p = \left[\sum_i \left(J_i - J_i^*\right)^p\right]^{1/p}, \quad p \geq 1, \tag{1}$$

where * indicates an optimal value. If $p = 1$ we have a linear combination, if $p = \infty$ we have the minimax method, and for $p = 2$ we minimize the Euclidean distance. This last choice constitutes Salukvadze's method [17], named after the Soviet scientist who originally proposed it. Weighting can be added in all cases.

The Salukvadze technique seems more attractive than the rest because of its simplicity and conceptual elegance. Furthermore, the minimization of distance from target quantities implies that there is a possibility to guide the overall system toward the direction that the designer considers optimal in some sense. This setting offers a large number of options since several requirements can be quantified and included in the performance index.

In a basic formulation the space of optimization is the two-dimensional target space of controller performance and estimator performance. The definition of appropriate corresponding targets, which we denote by J_c^* and J_e^* for control and estimation, respectively, determines the ideal point. Their selection is the topic of the next section. We can mention here the fact that they can be constant, time varying, or periodically updated, on line or off line. Therefore the multicriteria problem is to choose a control u such that

$$\min_u \|J - J^*\|^2 \tag{2}$$

is achieved, where

$$J = \begin{bmatrix} J_c \\ J_e \end{bmatrix}, \quad J^* = \begin{bmatrix} J_c^* \\ J_e^* \end{bmatrix}. \tag{3}$$

This can be written equivalently as

$$\min_u \left\{\left(J_c - J_c^*\right)^2 + \left(J_e - J_e^*\right)^2\right\}. \tag{4}$$

The control which at each step minimizes Eq. (2) is taken to be the (sub)optimal adaptive control for the interval [k, k + 1). Figure 1 presents the structure of the algorithm in flow diagram form. A closer look at this diagram will reveal that this

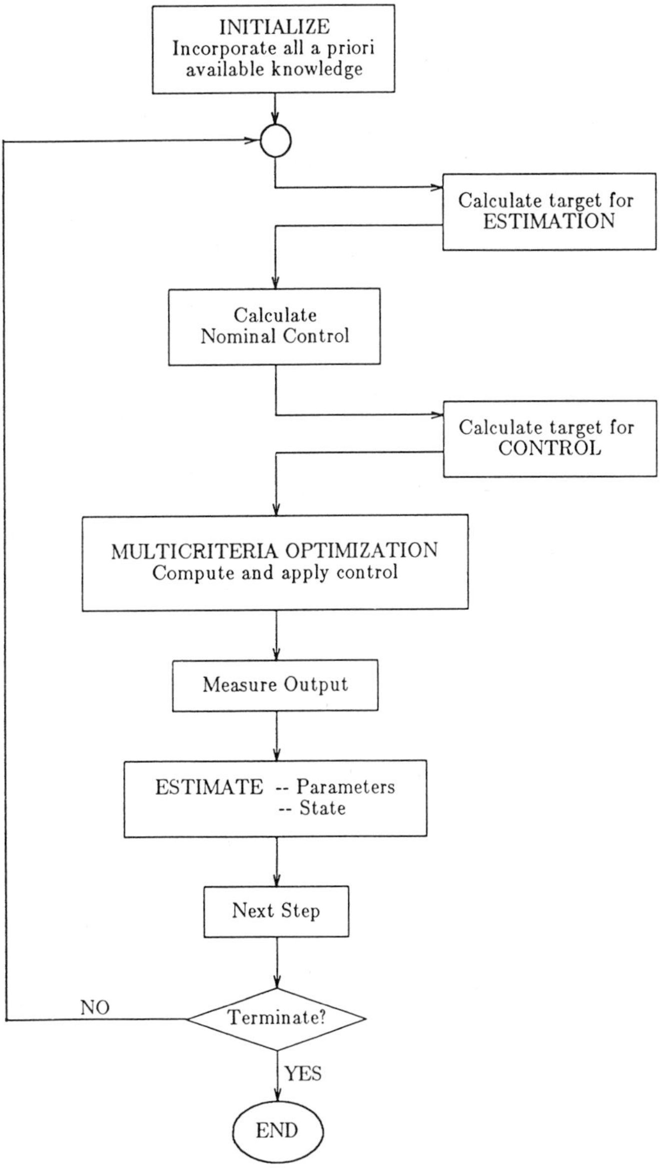

Fig. 1. New adaptive control algorithm.

is the basic functional strructure of most adaptive control algorithms (except the target evaluations, of course). We can thus see that these algorithms can be enriched, conditions permitting, with this multicriteria scheme. The next section presents the important point of target selection.

IV. TARGET SELECTION

The selection of the targets J_c^* and J_e^* depends on the particular algorithms chosen to perform the estimation and control tasks (see Fig. 1). The targets should be the goals that the estimator and controller procedures must reach, in the sense of the minimum distance from the ideal point which tese two targets define. It is evident that these quantities have the potential to determine, to a large degree, the behavior of the combined system, since excessive demands will cause large input signals which may finally result in instability. It is, therefore, important to select these targets in a proper manner.

There are several options in choosing the targets. We can have open-loop selection, feedback selection, or a combination of both, as well as on-line adjustments of any of the above. These terms are now explained. By open-loop selection we mean that without any information about the system, except that initially available, we arrive at specifying a target based on estimation and control theoretic facts, intuition, or experience (possibly from previous simulations or operation records). By feedback selection we mean that we use the information obtained at each step, as well as all past data, to derive the targets, using again theoretic facts or similar knowledge. In this class we can include a sort of closed-loop selection, in the spirit of [11], where the knowledge that the loop will remain closed in the future is used in order to anticipate the effects of actions in the present. In this last case a true closed-loop policy is realized. Moreover, by updating only the targets in closed loop, instead of using nominal trajectories as in [11], considerable savings in computation time and load can be achieved. Finally, the designer is free to make on-line adjustments by monitoring the targets' evolution in time, as well as the system performance, and deciding if they are satisfactory or not. These additional adjustments, however, may result in higher cost.

In most existing adaptive control schemes, the estimator represents the more complicated process. As if to counterbalance this fact, many estimation algorithms available today have been thoroughly studied and the factors affecting their performance are known and understood. The designer can thus explore this body of knowledge in order to obtain relations, mathematical or empirical, that can be used to select the estimation target and compute its numerical value. On the other hand, for the control target, the designer has again a number of options: linear-quadratic control theory, possibly combined with certainty equivalence or model following, can be used to define a nominal control. Moreover, the target need not be derived only through a performance index. It may be desirable to approximate a preselected type of control, as opposed to a control performance. In such a case we can use the MCO to force a quick convergence in the initial transition period. Thus, if the control target is the self-tuning control, then the new adaptive control will eventually approach the self-tuning one, as the accelerated parameter estimation is being accomplished.

We can see from the above that the designer's efforts will most likely concentrate on the estimator target since the control target can be easily derived. This last task can be also performed, in the most simple way, through extrapolating the system trajectory and evaluating the incurring cost, and operation which usually consists of rapidly performed iterations. In the rest of this section we shall demonstrate the selection of the estimation target when the estimator is the extended Kalman filter (EKF) or any other linearized version of the Kalman filter.

It would be ideal to drive the estimation error to zero, but this is not feasible. It is preferable instead to try to force the estimation error covariance toward attaining its minimum value, or at least being very close to it. The Cramer-Rao inequality [20, p. 412] can be used to provide an expression for the lower bound. In the case of the Kalman filter there are specific expressions which give the bounds for the error covariance [21, pp. 234-243]. More explicitly,

$$P_{k|k} \geq [O(k, k - N) + C^{-1}(k, k - N)]^{-1}, \quad k \geq N, \tag{5}$$

where

$$O(k, k - N) = \sum_{i=k-N}^{k} \Phi^T(i, k)H^T(i)V^{-1}(i)H(i)\Phi(i, k), \tag{6}$$

$$C(k, k - N) = \sum_{i=k-N}^{k} \Phi(k, i + 1)G(i)W(i + 1)G^T(i)\Phi^T(k, i + 1), \tag{7}$$

$$\Phi(i, k) = F^{i-k}, \quad \Phi(i, i) = \text{identity matrix}. \tag{8}$$

O and C are the stochastic observability and controllability matrices, respectively, and (·) denotes time dependence. Note that these matrices refer to the used model and not to the true system. So, they will be assigned their estimated values from the linearized parameters of the EKF, at each step k. If further suboptimality can be tolerated and if simulations do not indicate instability trends, this requirement can be relaxed and at each step the relations

$$O_k = F^T O_{k-1} F + H^T V^{-1} H, \tag{9}$$

$$C_k = F C_{k-1} F^T + G W G^T \tag{10}$$

can be used to update the target; note that this last option is possible only for time-invariant systems.

The designer need not be concerned about the presence of inverses, since P in Eq. (5) is always finite for the following reason. The series involved in that sum are of the form

$$\sum_{i=1}^{n} F^i (F^T)^i, \tag{11}$$

and we know [22] that if $||F|| < 1$, where $||F||$ is the norm of matrix F, this series converges. Supposing that F is stable, that is, its norm is less than 1; then, the series which define O and C are convergent. But then the inverse of C will grow without bound and so will the expression $O + C^{-1}$. However, the inverse of this expression will be finite and in extreme cases will have values in the neighborhood of zero. On the other hand, if we start with an unstable F, the diverging term will be O, and the sum $O + C^{-1}$ will again grow without bound but, still, the inverse of the sum will be finite. Thus, this estimation target will always exist for finite- or infinite-horizon problems and for stable or unstable system models.

V. PERFORMANCE ANALYSIS

Even though significant advances [23,24] in the area of stability and convergence analysis of adaptive control schemes have been recently made, Wittenmark's remark about simulation [25, p. 726] is still valid. Beyond the assumptions of correct order guessing and positive realness, the Martingale-based method [23] requires that the control be defined implicitly through the dot product of a regression vector and the parameter estimate, which looks, approximately, like a linear combination of past input-output data. On the other hand, the ordinary differential equation (ODE) approach [24], beyond the numerous assumptions including the boundedness of input and output signals, results in a complex nonlinear system of equations of order N^2, where N is the number of unknown parameters.

The algorithm presented in the previous sections, like most adaptive control algorithms with elements of duality, computes the control signal through the optimization of a functional. It is the complexity of this kind of feedback that prohibits an analytical investigation of the properties of these schemes. The only recourse is, therefore, simulation.

In practical applications, especially if the system order is low, an effort should be made to simplify, as much as possible, the multicriteria optimization task. Perhaps the gradient of the functional, with respect to control, can be written explicitly and its zeros calculated. The LISP-based symbol manipulation packages REDUCE (on IBM machines), Macsyma (or Vaxima on DEC machines), etc. can prove to be of great help. Again, however, the simulation will be the indispensable tool for a final evaluation.

We now present a number of examples that display the way the new algorithm works.

A. EXTENDED KALMAN FILTER

We consider the scalar linear discrete-time system

$$x_{k+1} = ax_k + bu_k + w_k, \tag{12}$$

$$z_k = x_k + v_k, \tag{13}$$

where b is an unknown constant and w and v are zero-mean Gaussian white uncorrelated noises, with covariance constants q_w and q_v, respectively. The initial state has a Gaussian distribution with mean x_0 and covariance M_0.

Augmenting the system state with the unknown parameter b written as a new state y_k, we obtain

$$x_{K+1} = zx_k + y_k u_k + w_k, \tag{14}$$

$$y_{k+1} = y_k, \tag{15}$$

and we thus arrive at a nonlinear stochastic system. If we select as estimator the extended Kalman filter, the relevant equations are [14]

$$\hat{x}_{k+1} = a\hat{x}_k + \hat{y}_k u_k + K_k(z_k - \hat{x}_k), \quad \hat{x}_0 = 0, \tag{16}$$

$$\hat{y}_{k+1} = \hat{y}_k + L_k(z_k - \hat{x}_k), \quad \hat{y}_0 = y_0, \tag{17}$$

with

$$K_k = (ap_{11_k} + u_k p_{2_k})/(p_{11_k} + q_v), \tag{18}$$

$$L_k = p_{2_k}/(p_{11_k} + q_v), \tag{19}$$

$$p_{11_{k+1}} = a^2 p_{11_k} + 2ap_{2_k}u_k + u_k^2 p_{22_k} - K_k^2(p_{11_k} + q_v) + q_w, \tag{20}$$

$$p_{2_{k+1}} = ap_{2_k} + u_k p_{22_k} - (ap_{11_k} + u_k p_{2_k}) p_{2_k} / (p_{11_k} + q_v), \tag{21}$$

$$p_{22_{k+1}} = p_{22_k} - p_{2_k}^2 / (p_{11_k} + q_v), \tag{22}$$

where we defined

$$P_k = \begin{bmatrix} p_{11_k} & p_{2_k} \\ p_{2_k} & p_{22_k} \end{bmatrix} \tag{23}$$

as the covariance of the augmented state estimate error. The estimator performance index is the trace of P, namely, $p_{11_k} + p_{22_k}$. We can use Eqs. (5)-(7) in order to calculate the estimation target; thus

$$est_k^t = c_k/(i_k c_k + 1), \tag{24}$$

where

$$c_k = a^2 c_{k-1} + q_w, \quad c_0 = q_w, \tag{25}$$

$$i_k = a^2 i_{k-1} + q_v, \quad i_0 = q_v. \tag{26}$$

For the control part we have several possibilities. We can choose the one-step-ahead quadratic cost or the general one over the period of operation. This, however, would significantly increase the complexity of the multicriteria global cost. Instead, we choose to approximate the enforced certainty-equivalent steady-state control, which is expressed as

$$u_k^t = \frac{-a\hat{x}_k\left(e_k + \sqrt{e_k^2 + 4QR\hat{b}_k^2}\right)}{\hat{b}_k\left(2r + e_k + \sqrt{e_k^2 + 4QR\hat{b}_k^2}\right)}, \tag{27}$$

where

$$e_k = Q\hat{b}_k^2 + (a^2 - 1)R, \tag{28}$$

Q and R being the weights in a LQG-type quadratic performance index. Thus, finally, the adaptive control will be derived through the minimization of the functional

$$J = \left(p_{11_{k+1}} + p_{22_{k+1}} - est_k^t\right)^2 + \left(u_k - u_k^t\right)^2. \tag{29}$$

Taking the derivative of J with respect to u, we obtain

$$2\left(p_{11_{k+1}} + p_{22_{k+1}} - est_k^t\right)\frac{\partial\left(p_{11_{k+1}} + p_{22_{k+1}}\right)}{\partial u_k} + 2\left(u_k - u_k^t\right) = 0. \tag{30}$$

Substituting Eqs. (20) and (22) into Eq. (30), we arrive at the cubic algebraic equation

$$\begin{aligned}&\left[p_{22_k}(p_{11_k} + q_v) - p_{2_k}^2\right]^2 u_k^3 + 3aq_v p_{2_k}\left[p_{22_k}(p_{11_k} + q_v) - p_{2_k}^2\right]u_k^2\\ &+ \Big\{(p_{11_k} + q_v)^2 + 2a^2 p_{2_k}^2 q_v^2 + \left[p_{22_k}(p_{11_k} + q_v) - p_{2_k}^2\right]\\ &\times \left[(p_{11_k} + q_v)\left(q_w + a^2 p_{11_k} + p_{22_k} - est_k^t\right) - p_{2_k}^2 - a^2 p_{11_k}\right]\Big\}u_k\\ &+ ap_{2_k} q_v \left[(p_{11_k} + q_v)\left(q_w + a^2 p_{11_k} + p_{22_k} - est_k^t\right) - p_{2_k}^2 - a^2 p_{11_1}\right]\\ &- u_k^t (p_{11_k} + q_v)^2 = 0,\end{aligned} \tag{31}$$

whose real solution with minimum norm is the optimal control for time k.

In order to avoid difficulties in the implementation of this scheme we have to take some measures:

(a) The coefficient of u_k^3 in Eq. (31) is going to appear as a denominator when we solve that cubic equation. As estimation improves, this term will become smaller and smaller, rendering the normalized coefficients larger and larger, thus creating numerical (overflow) problems. Therefore, when the estimation error goes below a prespecified low value, the adaptation part is switched off.

(b) The control signal u might assume exceedingly large values in the course of computation, possibly because of a situation as described in the preceding paragraph. A limiter must therefore be inserted to keep the control values within reasonable bounds.

We shall now present the results of simulations of the above scheme on a DEC VAX-11/750 computer with single-precision arithmetic. The pseudo-random noise generator used was approximately white (about 7% maximum correlation, instead of zero). The normal version of EKF was used, that is, without the gain sensitivity modification proposed in [14]. The horizon was taken to be 500 steps and the noise covariance constants were 0.1 for the plant and 1.0 for the measurements. The control limits were ±5.0 for the first four entries of Table 1 and ±10.0 for the last two.

The EKF converged quickly (within 40 steps at most) to a slightly biased estimate of b, and it managed to do so even from points significantly far from the neighborhood of the true value, thus circumventing a well-known limitation of "augmented" adaptive control schemes [10]. Furthermore, the estimation target converged to a value very close to the true value, that is, the one corresponding to known b. There have been strong indications that mismatch of the assumed and true covariances does not cause divergence of the algorithm, just additional, but still very small, bias.

TABLE 1.

Covariance limit	*Initial value*	*Final value*	*True value*
0.001	1.0 ± 1.0	0.676 ± 0.00091	0.6
0.001	1.0 ± 3.0	0.681 ± 0.00090	0.6
0.001	-1.0 ± 3.0	0.681 ± 0.00089	0.6
0.001	0.1 ± 1.0	0.583 ± 0.00094	0.6
0.001	0.1 ± 1.0	0.569 ± 0.00078	0.6
0.001	-1.5 ± 1.0	0.567 ± 0.00079	0.6

The EKF-based adaptive controllers are not widely in practical use. Probably, through the use of the multicriteria approach, this situation will change. The purpose of the above presentation was mainly to provide a demonstration for the implementation of the MCO-based adaptive control. The designer, through the limits on control signal and estimation error, has at his disposal the additional degrees of freedom needed to adjust or "tune" the scheme for any situation. In the next subsection we shall compare the new adaptive control scheme with the adaptive minimum-variance controller, which when combined with a least-squares estimator is usually known as the self-tuning controller.

B. *COMPARISON WITH THE SELF-TUNING REGULATOR*

As in the preceding subsection we consider a scalar system where, now, both parameters a and b are unknown but constant, the other conditions remaining the same. The control target is determined through the algebraic Riccati equation, whose solution is computed using the latest available estimates of the parameters involved. In other words we choose, as before, the enforced certainty-equivalent steady-state control. The estimation target, on the other hand, is not altered on line. Instead, a satisfactorily small value is specified beforehand as the target of the corresponding entry of the error covariance matrix. Thus, we have a vector target instead of a single value.

To assess the quality of adaptation in a succinct way, we follow the method of [26], where the covariance (or the autocorrelation) of the signals of interest is used to reveal the needed information. It is mentioned again here that the lower the initial covariance value and the faster the remaining coefficients drop near zero, the better the adaptive controller; furthermore, oscillations of the covariance values indicate an oscillatory signal.

The simulation results with measurement noise covariance 0.45 and estimation targets 0.05 for both parameters have been summarized in Table 2. The perturbation entries indicate the presence or absence of an additional excitation signal with (mean) amplitude and covariance as appearing in the parentheses. The autocorrelation functions of the output signals of cases 1, 2, 4, and 8 in Table 2 are plotted in Figs. 2 to 5.

From these data we can get an idea of how the multicriteria adaptive scheme behaves. First of all, the need for an overimposed excitation input is obvious. Furthermore, the parameters of this signal affect to a large degree the performance of the closed-loop system. They therefore constitute an additional set of parameters that may be used to "tune" the algorithm during the initial phase of the implementation, along with the targets' magnitudes and the initial conditions. The danger of overdriving the system is apparent from the last entry of Table 2, where the controller learned perfectly the

TABLE 2.

	Initial value	Final value	Perturbation	Cost
Self-tuning controller				
	a=0.1±7.0	$\hat{a}$=0.326±0.00225	No	203.9582
	a=0.1±7.0	$\hat{b}$=0.589±0.01105		
Multicriteria adaptive controller				
1	a=0.1±7.0	$\hat{a}$=0.356±0.00432	No	16.8512
	b=0.1±7.0	$\hat{b}$=0.323±4.47803		
2	a=0.1±7.0	$\hat{a}$=0.339±0.00376	Yes (0.01,0.55)	131.9083
	b=0.1±7.0	$\hat{b}$=0.636±0.01993		
3	a=0.1±7.0	$\hat{a}$=0.365±0.00411	Yes (0.05,0.55)	201.4940
	b=0.1±7.0	$\hat{b}$=0.421±0.41622		
4	a=0.1±7.0	$\hat{a}$=0.357±0.00395	Yes (0.1,0.55)	201.6204
	b=0.1±7.0	$\hat{b}$=0.506±0.22549		
5	a=0.1±7.0	$\hat{a}$=0.343±0.00384	Yes (0.2,0.55)	201.8671
	b=0.1±7.0	$\hat{b}$=0.577±0.07781		
6	a=0.1±7.0	$\hat{a}$=0.338±0.00373	Yes (0.3,0.55)	202.0473
	b=0.1±7.0	$\hat{b}$=0.597±0.03593		
7	a=0.1±7.0	$\hat{a}$=0.334±0.00361	Yes (0.4,0.55)	202.1900
	b=0.1±7.0	$\hat{b}$=0.603±0.02040		
8	a=0.1±7.0	$\hat{a}$=0.300±0.00003	Yes (0.5,0.55)	12780.70
	b=0.1±7.0	$\hat{b}$=0.600±0.00001		

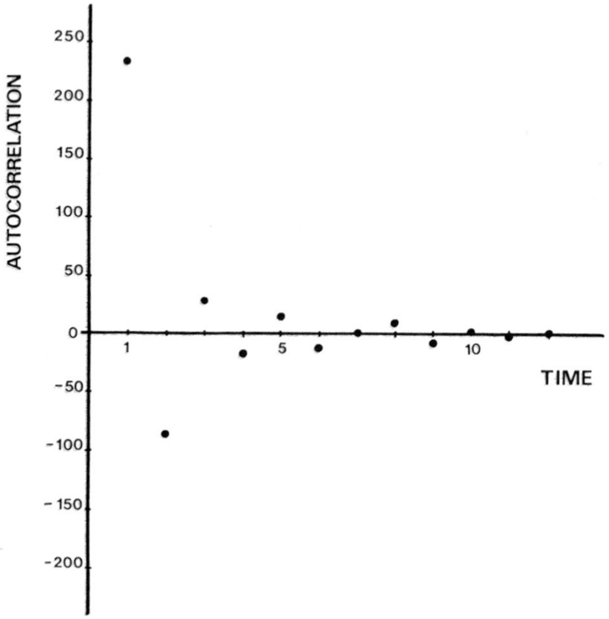

Fig. 2. Autocorrelation function of the output signal of case 1 in Table 2.

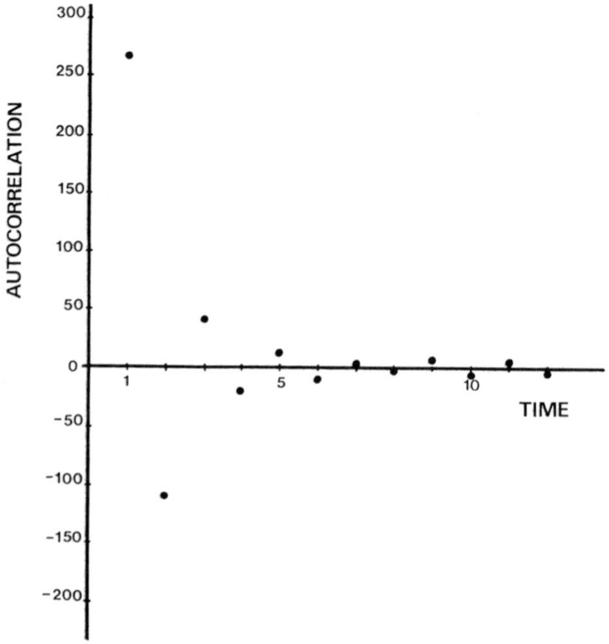

Fig. 3. Autocorrelation function of the output signal of case 2 in Table 2.

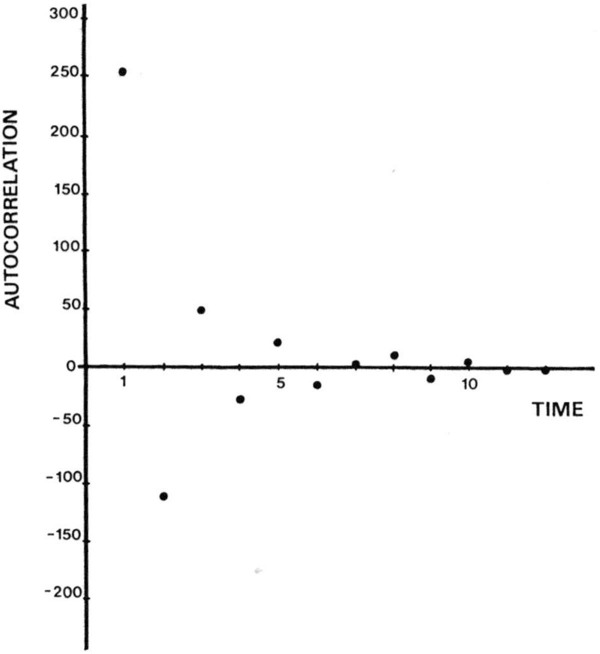

Fig. 4. Autocorrelation function of the output signal of case 4 in Table 2.

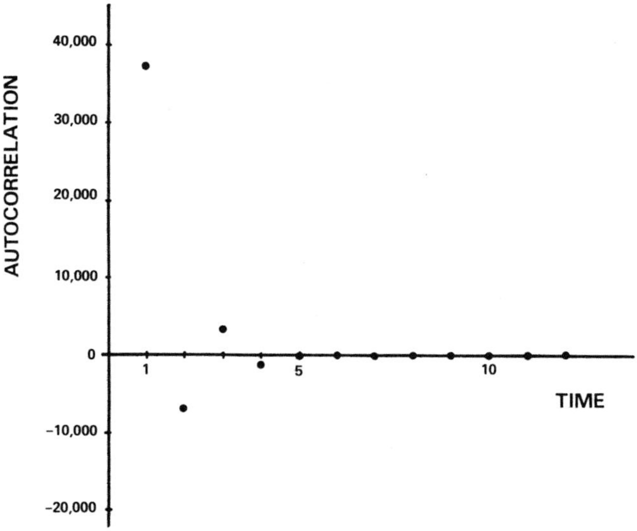

Fig. 5. Autocorrelation function of the output signal of case 8 in Table 2.

unknown parameters but at an exorbitant cost. On the other hand, one can see the merit of the multicriteria method by noticing that it can obtain comparable performance with the self-tuning controller at a lower cost, but at the price of increased implementation complexity.

VI. CONCLUSIONS

In this article we have presented a new approach to the adaptive and stochastic control problem. Aiming for an achievement of balance between estimation and control "expenditures," we exploited the conflict resolution capability of the Pareto-type multicriteria optimization. From the large number of approaches to this last problem, Salukvadze's ideal point method has been singled out for its simplicity and the chance it gives the designer to actually drive the closed-loop system toward specified goals which can be determined on or off line. This multicriteria optimization part can be incorporated in the structure of virtually every existing adaptive or stochastic control scheme without regard to the particular estimation/identification or control algorithm used. The new method makes possible the adaptive control of nonlinear systems as well. The price to be paid for such generality is the increased complexity of the controller and the lack of any guarantees for convergence and stability. Let us not forget, however, that stability and parameter convergence have been proven for a very small number of existing algorithms, mostly deterministic. In any case, extensive simulations will have to be performed before a practical implementation.

The new approach yielded good results for an adaptive control algorithm based on the extended Kalman filter. Even though only a scalar system has been considered in detail, there are strong indications that the multicriteria approach may be capable of rejuvenating this type of adaptive control. When it comes to linear systems, nothing can at present replace the adaptive minimum-variance controller, whose stability has been recently proven under restrictive but reasonable assumptions. The purpose of the relevant example in Section V,C was only to demonstrate the fact that the multicriteria adaptive controller can do the job.

Moreover, many interesting questions can be posed. For example, is it possible to obtain better results, foremostly, in the sense of decreased complexity by using another multicriteria optimization method or concept, especially by replacing Pareto-type solutions with minimax? Or, which existing adaptive control algorithm may benefit most by the introduction of the target concept? Finally, we may say that the value of the new approach, at least as an alternative, will probably be fully realized when conventional schemes are not able to perform well due to nonlinearities and/or violation of their inherent assumptions.

REFERENCES

1. T. W. KRAUS and T. J. MYRON, *Control Eng.*, 106 (1984).

2. J. ALSTER and P. R. BÉLANGER, *Automatica 10*, 627 (1974).

3. B. WITTENMARK, *Autom. Control Theory Appl. 3*, 13 (1975).

4. O. BADR, D. REY, and P. LADET, *IFAC Symp. Digital Comput. Appl. Process Control*, 141 (1977).

5. R. MILITO, C. S. PADILLA, R. A. PADILLA, and D. CADORIN, *Proc. IEEE Conf. Decision Control*, 341 (1980).

6. Y. HAIMES, "The Integration of System Identification and System Optimization," *Ph.D. Dissertation*, University of California, Los Angeles, 1970.

7. M. S. MAHMOUD, M. F. HASSAN, and M. G. SINGH, *Large Scale Syst. 1*, No. 3, 159 (1980).

8. R. H. LI and M. G. SINGH, *Large Scale Syst. 2*, 205 (1980).

9. P. D. ROBERTS and T. W. C. WILLIAMS, *Automatica 17*, No. 1, 199 (1981).

10. G. N. SARIDIS, "Self-Organizing Control of Stochastic Systems," Dekker, New York, 1977.

11. E. TSE and Y. BAR-SHALOM, *in* "Control and Dynamic Systems," Vol. 12, p. 99 (C. T. Leondes, ed.), Academic Press, New York, 1976.

12. R. ISERMANN, *Automatica 18*, No. 5, 513 (1982).

13. K. J. ASTRÖM, U. BORRISON, L. LJUNG, and B. WITTENMARK, *Automatica 13*, 457 (1977).

14. L. LJUNG, *IEEE Trans. Autom. Control AC-24*, 36 (1979).

15. L. LJUNG and T. SÖDERSTRÖM, "Theory and Practice of Recursive Identification," MIT Press, Cambridge, Massachusetts 1983.

16. V. CHANKONG and Y. Y. HAIMES, "Multiobjective Programming and Decision Making," North-Holland Publ., Amsterdam, 1983.

17. M. E. SALUKVADZE, "Vector-Valued Optimization Problems in Control Theory," Academic Press, New York, 1979.

18. N. T. KOUSSOULAS and C. T. LEONDES, *Int. J. Control 43*, No. 2, 337 (1986).

19. J. MEDANIĆ, *in* "Multicriteria Decision Making," p. 55, (G. Leitmann and A. Marzollo, eds.), Springer-Verlag, Berlin and New York, 1975.

20. P. EYKHOFF, "System Identification—Parameter and State Estimation," Wiley, New York, 1974.

21. A. H. JASWINSKI, "Stochastic Processes and Filtering Theory," Academic Press, New York, 1970.

22. A. S. HOUSEHOLDER, "The Theory of Matrices in Numerical Analysis," Ginn (Blaisdell), Boston, Massachusetts, 1964.

23. G. C. GOODWIN, P. J. RAMADGE, and P. E. CAINES, *SIAM J. Control Optim. 19*, 829 (1981).

24. L. LJUNG, *IEEE Trans. Autom. Control AC-22*, 551 (1977).

25. B. WITTENMARK, *Int. J. Control 21*, No. 5, 705 (1975).

26. M. FJELD and R. G. WILHELM, JR., *Control Eng.*, 99 (1981).

Instrumental Variable Methods for ARMA Models

PETRE STOICA

Facultatea de Automatica
Institutul Politehnic Bucuresti
Splaiul Independentei 313
R 77 206 Bucharest, Romania

BENJAMIN FRIEDLANDER

Saxpy Computer Corporation
255 San Geronimo Way
Sunnyvale, California 94086

TORSTEN SÖDERSTRÖM

Department of Automatic Control
and Systems Analysis
Institute of Technology
Uppsala University
S-751 21 Uppsala, Sweden

I. INTRODUCTION

The need for estimating the parameters of an autoregressive moving-average (ARMA) process arises in many applications in the areas of signal processing, spectral analysis, and system identification. A computationally attractive estimation procedure, which has received considerable attention in the literature, is based on a two-step approach: first, the autoregressive (AR) parameters are estimated using the modified Yule-Walker (MYW) equations; then, the moving average (MA) parameters are estimated by one of several available techniques.

In this article we consider only the first step of estimating the autoregressive parameters. In many engineering applications the second estimation step is not needed. The prime example is the estimation of autoregressive signals corrupted by white (or more generally, moving-average) measurement noise. In this case all the information about the spectral shape of the signal lies in the AR parameters of the signal-plus-noise ARMA process (see, e.g., [29]).

The relative simplicity of the MYW estimator motivated a number of authors to investigate this technique and to develop various extensions and variations [1-10]. Most of this work has been done in the context of high-resolution spectral analysis. One of the important observations made in these studies is that significant improvements in estimation accuracy can be obtained by increasing the number of MYW equations [2,9]. The resulting set of overdetermined equations is then solved by some least-squares technique. The possibility of using a weighted least-squares procedure was also discussed (see, e.g., [1,2]).

The performance evaluation of the MYW method was done mainly by simulation. In this article we provide an asymptotic accuracy analysis that clarifes the precise role of increasing the number of equations and of including a weighting matrix. It provides a valuable verification for experimental observations as well as guidelines for further improvements of MYW-based ARMA estimation techniques.

The MYW method is related to the instrumental variable (IV) method of parameter estimation [8,11,12]. In Section II we define an IV estimator that is slightly more general than the MYW estimators presented in the literature. In Section III we establish the consistency of the IV estimates and develop an explicit formula for the covariance matrix of the estimation errors. This formula can be used to evaluate the asymptotic performance of various MYW algorithms proposed in the literature [23,30]. In Section IV we study the optimization of estimation accuracy with respect to the weighting matrix and the number of equations. We show the existence of an optimal choice of the weighting matrix that minimizes the covariance matrix of the estimation errors. Furthermore, we show that the optimal error covariance matrix decreases monotonically when the number of equations is increased and converges as the number of equations tends to infinity. The form of this limiting matrix is also presented, and in Section V it is shown that it equals the asymptotic error covariance of the prediction error method. The effect of a certain filter used in the generation of the instrumental variables on the convergence rate of the error covariance matrix of the optimally weighted IV estimate is studied in Section VI. It is shown that there exists an optimal choice of this filter that gives the fastest convergence.

The main difficulty associated with the optimal IV method is that the optimal weighting matrix and the optimal prefilter depend on the second-order statistics of the data, which are not known a priori. In Sections VII and VIII we propose several multistep algorithms for overcoming this difficulty. Three approximate implementations of the OIV are presented and analyzed in Section VII: one based on an optimal weighting matrix and the other two on an optimal prefiltering operation. The implementation of three forms of the OIV estimator by means of a multistep procedure is discussed in Section VIII. The performance of the proposed estimation techniques is studied in Section IX by means of some numerical examples.

The work presented here provides an extension of IV methods that are usually applied to system identification to problems of time series analysis. For an overview of IV methods and their applications, see [11,12].

Finally, we note that results related to those presented here appeared recently in [31-33]. The problem considered in these references is the estimation of the parameters of dynamic econometric models by IV methods with instruments that are not exogenous. The approach used in [31-33] is based on a different formalism from the one used here.

II. THE ESTIMATION METHOD

Consider the following ARMA process of order (na, nc):

$$A(q^{-1})y(t) = C(q^{-1})e(t), \tag{1}$$

where e(t) is a white noise process with zero mean and variance λ^2,

$$A(q^{-1}) = 1 + a_1 q^{-1} + \cdots + a_{na} q^{-na},$$

$$C(q^{-1}) = 1 + c_1 q^{-1} + \cdots + c_{nc} q^{-nc},$$

and q^{-1} is the unit delay operator $[q^{-1}y(t) = y(t-1)]$. The following assumptions are made:

A1: $A(z) = 0 \Rightarrow |z| > 1$; $C(z) = 0 \Rightarrow |z| > 1$. In other words, the ARMA representation (1) is stable and invertible. This is not a restrictive assumption (cf. the spectral factorization theorem, e.g., (28)).

A2: $a_{na} \neq 0$, $c_{nc} \neq 0$, and $\{A(z), C(z)\}$ are coprime polynomials. In other words, (na, nc) are the minimal orders of the ARMA model (1).

Next we introduce the notation

$$\phi(t) = [-y(t-1), \ldots, -y(t-na)]^T,$$
$$\theta = [a_1, \ldots, a_{na}]^T, \tag{2}$$
$$v(t) = C(q^{-1})e(t).$$

Then (1) can be rewritten as

$$y(t) = \phi^T(t)\theta + v(t). \tag{3}$$

The unknown parameter vector θ will be estimated by minimizing a quadratic cost function involving the data vector $\phi(t)$ and an IV vector $z(t)$:

$$\hat{\theta} = \arg\min_{\theta} \left\| \sum_{t=1}^{N} z(t)\phi^T(t)\ \theta - \sum_{t=1}^{N} z(t)y(t) \right\|_Q^2, \tag{4}$$

where N is the number of data points, $\|x\|_Q^2 \triangleq x^TQx$, A is a positive-definite matrix, and

$$z(t) = G(q^{-1})\begin{bmatrix} y(t-nc-1) \\ \vdots \\ y(t-nc-m) \end{bmatrix}, \qquad m \geq na, \tag{5}$$

where $G(q^{-1})$ is a rational filter. We assume that

A3: $G(q^{-1})$ is stable and invertible, and $G(0) = 1$.

It is straightforward to show that the IV estimate in (4) can be obtained by a least-squares solution of the following system of linear equations:

$$Q^{1/2}\left[\sum_{t=1}^{N} z(t)\phi^T(t)\right]\hat{\theta} = Q^{1/2}\left[\sum_{t=1}^{N} z(t)y(t)\right]. \tag{6}$$

The solution can be written explicitly as

$$\hat{\theta} = \left[R_N^T Q R_N\right]^{-1}\left\{R_N^T Q\left[\frac{1}{N}\sum_{t=1}^{N} z(t)y(t)\right]\right\}, \tag{7}$$

where

$$R_N \triangleq \frac{1}{N} \sum_{t=1}^{N} z(t)\phi^T(t) \tag{8}$$

and $R_N^T Q R_N$ is assumed to be invertible. Evaluating $\hat{\theta}$ via (7) is known to be numerically less stable than direct solution of (6), using, for example, the QR algorithm [25]. However, the explicit formula in (7) is useful for theoretical analysis.

It is worthwhile to note that (6) reduces to various AR estimation techniques that have been proposed in the literature: (i) for $G(z) = 1$ and $m = na$ (Q is irrelevant in this case) we obtain the basic modified Yule-Walker equations [7,17]; (ii) for $G(z) = 1$, $m > na$, and $Q = I$, we obtain the overdetermined Yule-Walker equations [1,2]; (iii) for $G(z) = 1$, $m > na$, and $Q \neq I$, we obtain the weighted form of the overdetermined Yule-Walker equations [5]. Strictly speaking, the equivalence of (6) to the techniques mentioned here is only asymptotic. If, for example, the unbiased sample covariances are used, the equations arising in these techniques cannot be put exactly in the form of an IV method. However, the (asymptotic) results derived in this article will still apply since the different definitions of the sample covariances are asymptotically equivalent.

III. CONSISTENCY AND ACCURACY

A. *CONSISTENCY*

It is well known that the IV estimate (7) is consistent under the following conditions:

$$\text{rank } R = na, \quad \text{where} \quad R = \lim_{N\to\infty} R_N, \tag{9}$$

$$\lim_{N\to\infty} \frac{1}{N} \sum_{t=1}^{N} z(t)v(t) = 0. \tag{10}$$

See [11-13] for proofs and for further discussion. Under assumptions A1-A3 the limits in (9) and (10) equal the expected values of the corresponding variables (see e.g., [15]). Condition (10) is, therefore, satisfied since z(t) and v(t) are clearly uncorrelated. For G(z) = 1 it is known that (9) is satisfied; see [10] and [14] for two different proofs. Next we show that the same is true for $G(z) \neq 1$.

Lemma 1

Under conditions A1-A3,

$$\text{rank } R = na, \quad \text{where} \quad R = E\{z(t)\phi^T(t)\}. \tag{11}$$

Proof. Let

$$x = [x_1 \cdots x_{na}]^T, \qquad X(z) = \sum_{i=1}^{na} x_i z_i. \tag{12}$$

We have the following obvious equivalences:

$$\begin{aligned} Rx = 0 &\Leftrightarrow E\{G(q^{-1})y(t - nc - j) \cdot X(q^{-1})y(t)\} = 0, \\ &\qquad j = 1, \ldots, m, \\ &\Leftrightarrow \frac{1}{2\pi i} \oint G(z)\frac{C(z)}{A(z)}\, z^{nc+j} X(z^{-1})\frac{C(z^{-1})}{A(z^{-1})}\, \frac{dz}{z} = 0, \\ &\qquad j = 1, \ldots, m, \\ &\Leftrightarrow \frac{1}{2\pi i} \oint G(z)\frac{C(z)\,[z^{na}X(z^{-1})]\,[z^{nc}C(z^{-1})]}{A(z)\,[z^{na}A(z^{-1})]}\, z^{j-1}\, dz = 0, \\ &\qquad j = 1, \ldots, m \quad (m \geq na). \end{aligned} \tag{13}$$

Now it follows from [16, lemma 1] that the integrand in (13), excluding the factor z^{j-1}, must be analytic inside the unit circle. There are na poles within the unit circle that should be canceled by the zeros of $z^{na}X(z^{-1})$. However, since $z^{na}X(z^{-1})$

has degree na - 1, the only way the integral can equal zero is if $X(z) \equiv 0$. Thus, the only solution of the equation $Rx = 0$ is $x = 0$, from which it follows that R is of full rank.

Since both conditions (9) and (10) hold, the IV estimate in (7) is consistent.

B. ACCURACY

The main topic of interest in this section is the analysis of the accuracy of the IV estimates obtained via (6). The results presented here are based on a general analysis of the IV method that was presented in [11-13] and, in particular, on the following theorem.

Theorem 1 [12,13]

Under conditions A1-A3, the normalized IV estimation error is asymptotically normally distributed,

$$\frac{\sqrt{N}}{\lambda}(\hat{\theta} - \theta) \xrightarrow[N \to \infty]{\text{distribution}} \mathcal{N}(0, P), \tag{14}$$

with zero mean and covariance matrix

$$P = (R^T QR)^{-1} R^T QSQR(R^T QR)^{-1}, \tag{15}$$

where

$$S = E\{[C(q^{-1})z(t)][C(q^{-1})z(t)]^T\}, \tag{16}$$

with $\hat{\theta}$, R defined by (6) and (11).

Theorem 1 can be used to evaluate various choices of Q, G(•), and m by comparing the accuracies of the resulting estimates. A detailed discussion of the optimization of estimator accuracy will be given in Sections IV-VI. Here we present some examples illustrating the effect of Q and m on estimation accuracy, to motivate the subsequent discussion.

Example 1: Influence of Q *and* m *on Accuracy*

Consider the following ARMA (1, 1) process:

$$(1 - 0.8q^{-1})y(t) = (1 + 0.7q^{-1})e(t), \quad E\{e(t)e(s)\} = \delta_{ts}. \tag{17}$$

Let z(t) be given by (5) with $G(q^{-1}) \equiv 1$ and $m \geq 1$. We investigated the influence of m and Q on estimation accuracy by evaluating the variance of the estimation error, P (15), for values of m between 1 and 10, and for the following weighting matrices Q:

$$Q_1 = I,$$

$$Q_2 = \begin{bmatrix} 2 & -1 & & 0 \\ -1 & 2 & & \\ & \ddots & \ddots & -1 \\ 0 & & -1 & 2 \end{bmatrix} \tag{18}$$

$$Q_3 = \begin{bmatrix} 1 & \rho & \cdots & \rho^{m-1} \\ \rho & \ddots & \ddots & \vdots \\ \vdots & \ddots & & \rho \\ \rho^{m-1} & \cdots & \rho & 1 \end{bmatrix} \quad \text{with} \quad \rho = 0.9 \quad \text{and} \quad \rho = 0.5.$$

The choice $Q_1 = I$ corresponds to the commonly used nonweighted modified Yule-Walker equations. The matrix Q_2 corresponds to a differencing operation on the elements of (6) since

$$Q_2^{1/2} \cong \begin{bmatrix} 1 & & & 0 \\ -1 & 1 & & \\ & \ddots & \ddots & \\ 0 & & -1 & 1 \end{bmatrix}. \tag{19}$$

The matrix Q_3 is the autocorrelation matrix of a first-order AR process with a pole at $z = \rho$. See [5] for other choices of the weighting matrix that are useful in some applications.

The results are summarized in Table 1. Note that the accuracy does not necessarily increase with increasing m. This is in contrast with the implicit assumption in the literature on the overdetermined MYW equations that increasing the number of equations always improves the estimates [2,3]. Furthermore, it appears difficult, if not impossible, to predict which *ad hoc* weighting matrix will lead to best accuracy.

We have also compared the accuracy of the IV estimate with that given by the prediction error method (PEM) [18,19,26], for some simple low-order systems (see, e.g., Example 2 (Section V) and the examples in [23]). Recall that in the Gaussian case, the PEM error covariance matrix equals the Cramer-Rao lower bound. The differences in accuracy between the IV method and

Table 1. Influence of m and Q on the Accuracy of the IV Estimate (6)

	P			
m	$Q = Q_1$	$Q = Q_2$	$Q = Q_3$ ($\rho = 0.9$)	$Q = Q_3$ ($\rho = 0.5$)
1	0.471	0.471	0.471	0.471
2	0.546	0.523	0.559	0.555
3	0.612	0.526	0.649	0.637
4	0.668	0.512	0.742	0.715
5	0.715	0.496	0.839	0.788
6	0.754	0.484	0.939	0.852
7	0.786	0.475	1.040	0.907
8	0.810	0.469	1.141	0.953
9	0.829	0.466	1.239	0.984
10	0.844	0.464	1.335	1.018

the PEM were sometimes considerable, indicating that the IV estimator with *ad hoc* choices of Q, m, and $G(q^{-1})$ is inefficient (in the statistical sense).

The questions raised above motivate more detailed examination of the accuracy aspects of the IV estimates. In particular, it is of interest to choose Q, m, and $G(q^{-1})$ so as to increase the accuracy of the IV estimate (6). This is discussed in Sections IV-VI.

IV. OPTIMIZATION OF ESTIMATION ACCURACY

The problem of determining optimal IV estimates in the fairly general class of estimates defined by (6) can be stated as follows: find Q_{opt}, m_{opt}, and $G_{opt}(q^{-1})$ such that the corresponding covariance matrix P_{opt} has the property $P \geq P_{opt}$, where P corresponds to any other admissible choice of Q, m, and G. This type of problem was studied in [12,13] for systems with exogenous inputs, such as ARMAX systems. The results of [12,13] cannot be applied directly to the ARMA problem, as is explained in [30]. Therefore, we must approach the accuracy optimization in another way. As we will see, the optimization with respect to Q, m, and $G(q^{-1})$ can be treated in three distinct steps. We start with the optimization of P given by (15) and (16) with respect to the weighting matrix Q, for which the following result holds.

Theorem 2

Consider the matrix P defined in (15). We have

$$P \geq (R^T S^{-1} R)^{-1} \triangleq \tilde{P}_m. \tag{20}$$

Furthermore, the quality $P = \tilde{P}_m$ holds if and only if

$$SQR = R(R^TS^{-1}R)^{-1}(R^TQR). \tag{21}$$

Proof. It is straightforward to show that

$$P - \tilde{P}_m = [(R^TQR)^{-1}R^TQ - (R^TS^{-1}R)^{-1}] \times S[(R^TQR)^{-1}R^TQ - (R^TS^{-1}R)^{-1}R^TS^{-1}]^T. \tag{22}$$

Since $S > 0$, (20) and (21) follow.

Note that (20) is closely related to the Gauss-Markov theorem in regression theory [22]. An obvious way to satisfy (21) is to set $Q = S^{-1}$, in which case $P = \tilde{P}_m$.

Next we consider the optimization of $\tilde{P}_m$ with respect to m. In Section V (Lemma 2) we will formally prove that for the optimal choice of Q, estimation accuracy increases monotonically with m, that is, $\tilde{P}_m \geq \tilde{P}_{m+1}$ for all $m \geq na$. As was shown earlier, this is not true for arbitrary choices of Q (see also [23,30]).

Note that the above results are valid for general IV estimation problems. The detailed structure of the matrices R and S is not used anywhere in the proofs. Note also that for AR systems it can be shown that $\tilde{P}_{m+1} = \tilde{P}_m (m \geq na)$. However, for ARMA processes we have in general, $\tilde{P}_m > \tilde{P}_{m+1}$.

Since $\tilde{P}_m$ is monotonically decreasing and also $\tilde{P}_m > 0$, it follows that $\tilde{P}_m$ will converge to a limit as m tends to infinity. A formal discussion of the convergence of $\tilde{P}_m$ is given in Appendix B, where it is also shown that

$$\tilde{P}_\infty = \lim_{m\to\infty} \tilde{P}_m = \lambda^2 [E\{\phi(t)\psi^T(t)\}E\{\psi(t)\phi^T(t)\}]^{-1}, \tag{23}$$

where $\psi(t)$ is the following infinite-dimensional vector:

$$\psi(t) = \frac{1}{C(q^{-1})}\begin{bmatrix} e(t - nc - 1) \\ e(t - nc - 2) \\ \vdots \end{bmatrix}. \tag{24}$$

The limiting error covariance matrix $\tilde{P}_\infty$ can be evaluated by solving a certain discrete Lyapunov equation [see (A.5) and (B.17)]. Note that $\tilde{P}_\infty$ is independent of $G(\cdot)$. We will show, however, in Section VI that the choice of $G(\cdot)$ affects the "convergence rate" of $\tilde{P}_m$.

V. COMPARISON OF THE ACCURACIES OF THE OPTIMAL IV METHOD AND THE PREDICTION ERROR METHOD

The prediction error method has been studied widely in the context of system identification [18,19,26]. The prediction error estimate of the parameters $\{a_i, c_i\}$ of an ARMA process is obtained by minimizing the loss function

$$V_N\left(\hat{a}_1, \ldots, \hat{a}_{na}, \hat{c}_1, \ldots, \hat{c}_{nc}\right) = \sum_{t=1}^{N} \epsilon^2(t), \tag{25}$$

where

$$\epsilon(t) = \frac{\hat{A}(q^{-1})}{\hat{C}(q^{-1})} y(t). \tag{26}$$

The prediction error estimate is known to be asymptotically normally distributed with the following normalized covariance matrix:

$$\lim_{N\to\infty} \frac{N}{\lambda^2} \operatorname{cov}\left\{\hat{a}_1, \ldots, \hat{a}_{na}, \hat{c}_1, \ldots, \hat{c}_{nc}\right\}$$

$$= \left[E\left\{\begin{bmatrix} \psi_1(t) \\ -\psi_2(t) \end{bmatrix} \left[\psi_1^T(t), -\psi_2^T(t)\right]\right\}\right]^{-1}, \tag{27}$$

where

$$\psi_1^T(t) = \frac{1}{A(q^{-1})}[e(t-1), \ldots, e(t-na)], \tag{28}$$

$$\psi_2^T(t) = \frac{1}{C(q^{-1})}[e(t-1), \ldots, e(t-nc)]. \tag{29}$$

It is straightforward to show from (27) that the normalized covariance matrix of the AR parameter estimates obtained by the PEM is given by

$$P_{PEM} \triangleq \lim_{N\to\infty} \frac{N}{\lambda^2} \operatorname{cov}\{\hat{\theta}\} = \left[D_{11} - D_{12}D_{22}^{-1}D_{12}^T\right]^{-1}, \tag{30}$$

where

$$\hat{\theta} = \left[\hat{a}_1, \ldots, \hat{a}_{na}\right]^T, \tag{31a}$$

$$D_{ij} = E\left\{\psi_i(t)\psi_j^T(t)\right\}, \quad i, j = 1, 2. \tag{31b}$$

The following result states the the optimal IV method has the same asymptotic accuracy as the PEM.

Theorem 3

Let $\tilde{P}_\infty$ and P_{PEM} be the covariance matrices defined by (23) and (28)-(31), respectively. Then, under assumptions A1-A3, $\tilde{P}_\infty = P_{PEM}$.

Proof. See Appendix A.

As was mentioned earlier, in the Gaussian case, the PEM is an efficient estimator, that is, P_{PEM} equals the Cramer-Rao lower bound [19,22]. We conclude, therefore, that the optimal IV method is an efficient estimator for Gaussian processes. If the data are not Gaussian, then the optimal IV estimate, like the PE estimate, will still give the minimum variance in the fairly large class of parameter estimators whose covariance matrices depend only on the second-order statistics of the data.

It is interesting to investigate the rate at which $\hat{P}_m$ converges to $\tilde{P}_\infty = P_{PEM}$, since in practice the value of m cannot be too large. The convergence rate of $\tilde{P}_m$ is illustrated by the following examples.

Example 2: Convergence of $\tilde{P}_m$ *to* P_{PEM}

Consider the ARMA processes

$$S_1: (1 - 0.8q^{-1})y(t) = (1 + 0.7q^{-1})e(t) \tag{32}$$

and

$$S_2: (1 - 1.5q^{-1} + 0.7q^{-2})y(t) = (1 - q^{-1} + 0.2q^{-2})e(t), \tag{33}$$

where in both cases $E\{e(t)e(s)\} = \delta_{t,s}$ $(\lambda^2 = 1)$. For both S_1 and S_2 we evaluated P_{PEM} and the optimal covariance matrix $\tilde{P}_m$, for $G(z) \equiv 1$ and $m = na, na + 1, \ldots$. The results are shown in Table 2, where $\tilde{P}_m^{ij}$ denotes the (ij)th element of $\tilde{P}_m$. Note

Table 2. Convergence of $\tilde{P}_m$ *to* $\tilde{P}_\infty$ *for* S_1 *and* S_2

	S_1	S_2		
m	$\tilde{P}_m^{11}$	$\tilde{P}_m^{11}$	$\tilde{P}_m^{12}$	$\tilde{P}_m^{22}$
1	0.471	--	--	--
2	0.426	52.190	-16.320	6.276
3	0.409	8.577	-2.181	1.689
4	0.401	3.298	-1.151	1.488
5	0.397	2.147	-1.093	1.485
6	0.394	1.804	-1.131	1.481
8	0.391	1.626	-1.189	1.461
10	0.390	1.589	-1.231	1.445
12	0.390	1.580	-1.222	1.436
14	0.389	1.577	-1.226	1.431
16	0.389	1.576	-1.227	1.429
18	0.389	1.576	-1.227	1.429
20	0.389	1.576	-1.227	1.429
P_{PEM}^{ij}	0.389	1.576	-1.227	1.429

that $\tilde{P}_m$ has essentially converged for m = 15. It is interesting to compare the accuracy of the optimal IV method to that of the basic modified Yule-Walker method (m = na, in which case the choice of Q is irrelevant). The difference in accuracies can be quite large. For example, in the case of S_2, the ratio of the variances of $\hat{a}_1$ corresponding to the two methods is about 30. For higher order systems the difference of accuracy between the methods may be larger. Consider the following fourth-order system:

$$S_3: (1 - 2.6q^{-1} + 3.58q^{-2} - 2.574q^{-3} + 0.9801q^{-4})y(t)$$
$$= (1 - 2.5q^{-1} + 3.5q^{-2} - 2.4q^{-3} + 0.8q^{-4})e(t),$$
$$E\{e^2(t)\} = 1. \qquad (34)$$

Table 3. Convergence of the Diagonal Elements of $\tilde{P}_m$ for S_3

	S_3			
m	$\tilde{P}_m^{11}$	$\tilde{P}_m^{22}$	$\tilde{P}_m^{33}$	$\tilde{P}_m^{44}$
4	789.700	3206.00	2180.00	348.700
5	42.230	10.85	54.79	5.653
6	2.543	3.24	1.29	0.216
7	0.148	0.71	0.91	0.216
8	0.084	0.64	0.62	0.111
10	0.058	0.50	0.23	0.046
12	0.048	0.22	0.21	0.045
14	0.045	0.20	0.20	0.045
16	0.044	0.20	0.20	0.044
18	0.044	0.19	0.19	0.044
20	0.043	0.19	0.19	0.043
P_{PEM}^{ii}	0.040	0.17	0.17	0.039

In Table 3 we show the variation of the diagonal elements $\tilde{P}_m^{ii}$ of $\tilde{P}_m$ with m increasing from 4 to 20. The values of P_{PEM}^{ii} are also shown for comparison. It can be seen that in the case of S_3 the ratio between the variances of the estimation errors corresponding to the basic MYW method (m = na) and the optimal IV method is $>10^4$.

Example 3: Convergence of $\tilde{P}_m$ *to* P_{PEM}

Note that $\tilde{P}_m$ approaches P_{PEM} more or less at an exponential rate (cf. Example 2). To investigate the convergence rate in more detail, consider the general ARMA (1, 1) process

$$y(t) = -ay(t-1) + e(t) + ce(t-1). \tag{35}$$

Assuming that

$$\tilde{P}_m = P_{PEM} + K\gamma^m, \quad 0 \leq \gamma < 1, \quad K = \text{constant}, \tag{36}$$

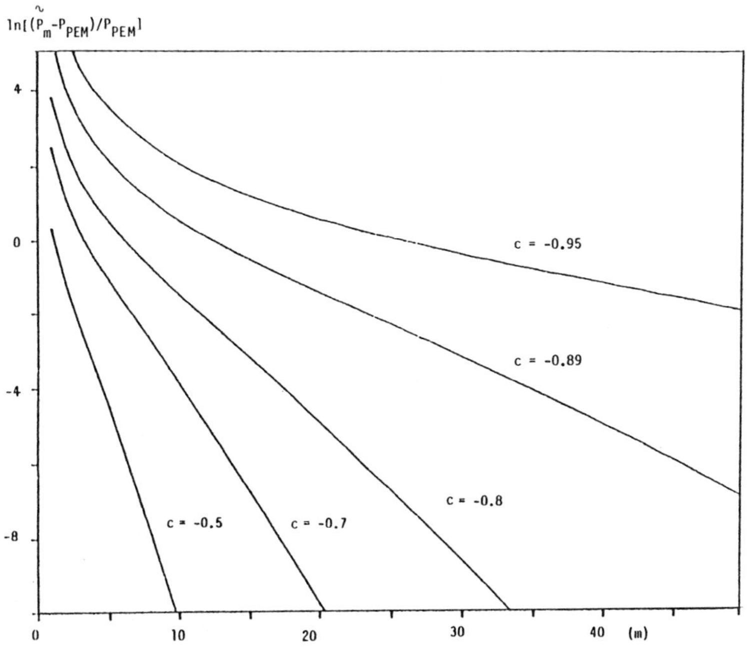

Fig. 1. Convergence of $\tilde{P}_m$*,* $a = -0.9$*, varying c.*

it seems reasonable to plot $\ln\left[\left(\tilde{P}_m - P_{PEM}\right)/P_{PEM}\right]$ versus m. This is done in Figs. 1 and 2 for $G(q^{-1}) \equiv 1$ and different values of the parameters a and c. It can be seen that except for small values of m, the curves can be well approximated by straight lines. This justifies the assumption in (36). It is interesting to note that the convergence rate depends strongly on c and only weakly on a. The convergence is particularly slow when c is close to -1 (zero near the unit circle). The large variations in convergence rates for different parameters of the data motivate the study of ways for improving the convergence rate. In the next section we show how the choice of $G(q^{-1})$ affects the convergence rate.

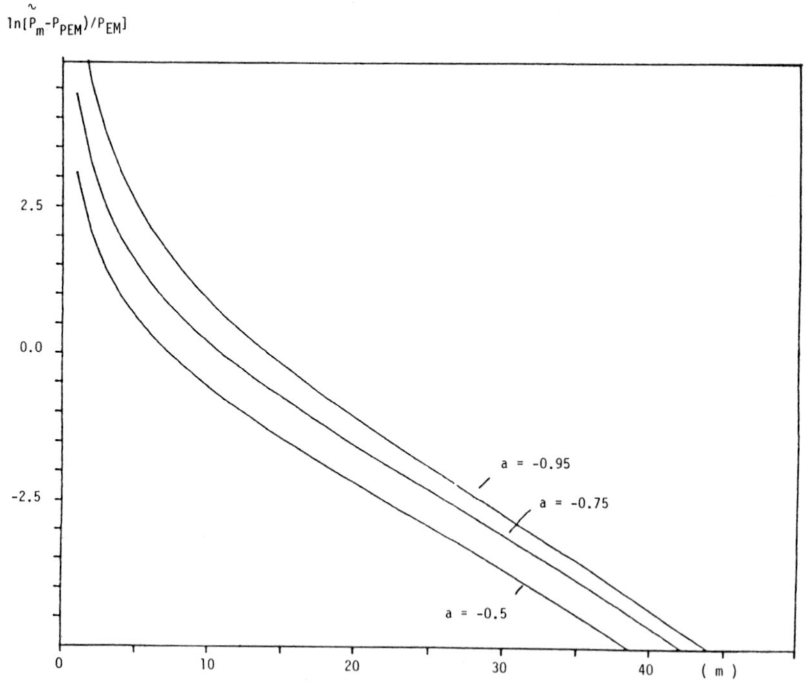

Fig. 2. Convergence of $\tilde{P}_m$, c = -0.9, varying a.

VI. THE OPTIMAL CHOICE OF $G(q^{-1})$

In this section we show that the choice $G(q^{-1}) = 1/C^2(q^{-1})$ will ensure that optimal estimation accuracy is achieved for a finite m, in fact, for $m = na$. To see this we state the following lemma. Note that in the following calculations we will add the subscript m to R, S, z(t), etc. to emphasize their dependence on the number of instrumental variables.

Lemma 2

The matrices $\{\tilde{P}_m\}$ form a nonincreasing sequence, that is, $\tilde{P}_{na} \geq \tilde{P}_{na+1} \geq \cdots \geq \tilde{P}_\infty$. Furthermore, all the equalities hold if and only if

$$R_m^T S_m^{-1} x_m = 0 \quad \text{for} \quad m \geq na, \tag{37}$$

where R_m, S_m are as defined by (11), (16), and

$$x_m \triangleq E\left\{C^2(q^{-1})G(q^{-1}) \times \begin{bmatrix} e(t-1) \\ \vdots \\ e(t-m) \end{bmatrix} \frac{C^2(q^{-1})G(q^{-1})}{A(q^{-1})} e(t-m-1)\right\}. \tag{38}$$

Proof. See Appendix C.

It is now easy to see that the choice $G(q^{-1}) = 1/C^2(q^{-1})$ will satisfy (37) and is, therefore, optimal (although not necessarily the only optimal choice). We state this formally in the following theorem.

Theorem 4

Let assumptions A1-A3 hold true and consider the IV estimate (6) with $m = na$ and $G(q^{-1}) = 1/C^2(q^{-1})$ (the choice of Q is irrelevant in this case). Under these conditions the IV estimate will be optimal in the sense that its asymptotic ($N \to \infty$) covariance matrix equals $\tilde{P}_\infty$ ($=P_{PEM}$).

Proof. Direct consequence of Lemma 2.

Another interesting choice for $G(q^{-1})$ is $G(q^{-1}) = A(q^{-1})/C^2(q^{-1})$. In this case, $S_m = \lambda^2 I_m$. Thus the optimal weighting matrix is $Q = I$ (the scaling factor $1/\lambda^2$ does not matter). Note that for this choice of $G(q^{-1})$ the equation (1) does not necessarily hold. Therefore, the covariance matrix of the corresponding IV estimator will approach P_{PEM} as $m \to \infty$. That is to say, unlike the case where $G(q^{-1}) = 1/C^2(q^{-1})$, when $G(q^{-1}) = A(q^{-1})/C^2(q^{-1})$, the matrix $\tilde{P}_m$ does not equal $\tilde{P}_\infty$ for $m = na$.

VII. THE OPTIMAL IV MULTISTEP ESTIMATES AND THEIR ASYMPTOTIC PROPERTIES

From the discussions in Sections IV and VI, it follows that we have (at least) three ways of generating optimal IV estimates using (6):

$$\text{OIV-1:} \quad Q = S_m^{-1}, \quad G(q^{-1}) = 1, \quad m \to \infty, \tag{39}$$

$$\text{OIV-2:} \quad q = I, \quad G(q^{-1}) = 1/C^2(q^{-1}), \quad m = na, \tag{40}$$

$$\text{OIV-3:} \quad Q = I, \quad G(q^{-1}) = A(q^{-1})/C^2(q^{-1}), \quad m \to \infty. \tag{41}$$

The problem is that all of these methods depend on knowledge of unknown quantities. This is the usual dilemma in accuracy optimization. Our aim here is to show how to overcome this difficulty for the case under consideration.

We will start by showing that replacing S_m (in OIV-1), $C(q^{-1})$ (in OIV-2), and $C(q^{-1})$, $A(q^{-1})$ (in OIV-3) by their consistent estimates will not affect asymptotic estimation accuracy. Then we show how to obtain such consistent estimates of S_m, $A(q^{-1})$, and $C(q^{-1})$. The proposed estimation procedures are therefore based on estimating S_m, $A(q^{-1})$, and $C(q^{-1})$ and using these estimates in (6) instead of the true S_m, $A(q^{-1})$, and

$C(q^{-1})$. As we will see, the implementation of OIV-1 does not require explicit computation of $C(q^{-1})$. This may be advantageous in applications where only the AR parameters need to be estimated. We analyze the asymptotic properties of OIV-1, OIV-2, and OIV-3 by techniques similar to those used in Section III and in [13].

A. *APPROXIMATE OIV-1*

Let $\hat{S}_m$ denote a consistent estimate of S_m. Let $\hat{\theta}_1$ denote the OIV-1 estimate (6), (39) for a given m [possibly $m = m(N)$, where m increases without bound as $N \to \infty$] and let $\hat{\hat{\theta}}$ be the approximate OIV-1 estimate with S_m replaced by $\hat{S}_m$. We will say that two consistent estimates of θ, say, $\hat{\theta}_1$ and $\hat{\hat{\theta}}_1$, are asymptotically equivalent if

$$\hat{\theta}_1 - \hat{\hat{\theta}}_1 = 0\left(|\hat{\theta}_1 - \theta|\right).$$

In other words, the difference $\left(\hat{\theta}_1 - \hat{\hat{\theta}}_1\right)$ tends to zero faster than the estimate error $\left(\hat{\theta}_1 - \theta\right)$, as N tends to infinity. We can now state the following theorem.

Theorem 5

Let assumptions A1-A3 be true, and assume also that $\hat{S}_m - S_m = 0(1/\sqrt{N})$ * and that $[m(N)]^8/N \to 0$ as $N \to \infty$. Then $\hat{\theta}_1$ and $\hat{\hat{\theta}}_1$ are asymptotically equivalent.

Proof. From the assumption above it follows that for sufficiently large N we have

$$\hat{S}_m^{-1} = (S_m + 0(1/\sqrt{N}))^{-1} = (I + 0(m/\sqrt{N}))^{-1}S_m^{-1}$$
$$= S_m^{-1} + 0(m^2/\sqrt{N}). \tag{42}$$

**We will use throughout the article $0(\epsilon)$ to denote a random variable with standard deviation $K\epsilon$, where ϵ is small and where K is a (finite) constant independent of ϵ.*

Note that A1-A3 imply that $\delta_1 I \le S_m \le \delta_2 I$, where $\delta_1 > 0$ and $\delta_2 < \infty$ do not depend on m. Note also that if m does not increase without bound as $N \to \infty$, then the terms $0(m^k/\sqrt{N})$ should be interpreted as $0(1/\sqrt{N})$. The OIV-1 solution of (6) can be written explicitly as

$$\hat{\theta}_1 = \left[R_N^T S_m^{-1} R_N\right]^{-1} \left\{ R_N^T S_m^{-1} \left[\frac{1}{N} \sum_{t=1}^{N} z_m(t) y(t) \right] \right\}, \tag{43}$$

where

$$R_N \triangleq \frac{1}{N} \sum_{t=1}^{N} z_m(t) \phi^T(t). \tag{44}$$

It is straightforward to show that

$$\hat{\theta}_1 - \theta = \left[R_N^T S_m^{-1} R_N\right]^{-1} \left\{ R_N^T S_m^{-1} \left[\frac{1}{N} \sum_{t=1}^{N} z_m(t) v(t) \right] \right\} = 0(m/\sqrt{N}). \tag{45}$$

Similarly,

$$\hat{\hat{\theta}}_1 - \theta = \left[R_N^T \hat{S}_m^{-1} R_N\right]^{-1} \left\{ R_N^T \hat{S}_m^{-1} \left[\frac{1}{N} \sum_{t=1}^{N} z_m(t) v(t) \right] \right\}. \tag{46}$$

From (42), (45), and (46) it follows that

$$\begin{aligned} \hat{\hat{\theta}}_1 - \theta &= \left\{ R_N^T \left[S_m^{-1} + 0(m^2/\sqrt{N}) \right] R_N \right\}^{-1} \\ &\quad \times \left\{ R_N^T \left[S_m^{-1} + 0(m^2/\sqrt{N}) \right] \frac{1}{N} \sum_{t=1}^{N} z_m(t) v(t) \right\} \\ &= \left\{ R_N^T S_m^{-1} R_N + 0(m^4/\sqrt{N}) \right\}^{-1} \\ &\quad \times \left[R_N^T S_m^{-1} \frac{1}{N} \sum_{t=1}^{N} z_m(t) v(t) + 0(m^4/N) \right] \\ &= \left(\hat{\theta}_1 - \theta \right) + 0(m^5/N). \end{aligned} \tag{47}$$

Since $m^5/N = (m/\sqrt{N})(m^4/\sqrt{N})$, and since $m^4/\sqrt{N} \to 0$ by assumption, it follows that the second term in (47) goes to zero as $N \to \infty$, faster than the term $\hat{\theta}_1 - \theta$ [which is $O(m/\sqrt{N})$, cf. (45)], which concludes the proof.

The convergence of $\tilde{P}_m$ to $\tilde{P}_\infty$ may be slow, especially if $C(z)$ has zeros close to the unit circle. For the idealized OIV-1 estimate $\hat{\theta}_1$, we may then need to consider a large m in order to obtain good accuracy. For the practical estimate $\hat{\hat{\theta}}_1$ the situation is, however, different. If m is too large with respect to N, then $\hat{\hat{\theta}}_1$ and $\hat{\theta}_1$ may not have the same distribution and thus $\hat{\hat{\theta}}_1$ may not be (asymptotically) optimal. Theorem 5 gives an upper bound on m [$m(N) = N^{1/(8-\delta)}$, $\delta >$)], guaranteeing that $\hat{\theta}_1$ and $\hat{\hat{\theta}}_1$ are asymptotically equivalent. However, no attempt has been made to give a tight bound. In fact, this seems quite difficult since a tight bound would be problem dependent. In Section IX we will discuss further this point and illustrate it by means of some simulations. It is shown there that the bound of Theorem 5 is quite conservative. That is to say, $\hat{\theta}_1$ and $\hat{\hat{\theta}}_1$ may behave similarly for $m \gg N^{1/8}$. As explained earlier, one needs to consider large values of m when the convergence of $\tilde{P}_m$ to $\tilde{P}_\infty$ is slow.

B. *APPROXIMATE OIV-2*

Let $\hat{C}(q^{-1})$ denote a consistent estimate of $C(q^{-1})$. Let $\hat{\hat{\theta}}_2$ be the approximate estimate obtained by replacing $C(q^{-1})$ by $\hat{C}(q^{-1})$ in the OIV-2 estimate $\hat{\theta}_2$ [(6), (40)]. Then we can state the following theorem.

Theorem 6

Let assumptions A1-A3 hold true and assume that $\hat{c}_i - c_i = 0(1/\sqrt{N})$, $i = 1, \ldots, nc$. Then $\hat{\theta}_2$ and $\hat{\hat{\theta}}_2$ are asymptotically equivalent.

Proof. The OIV-2 solution of (6) can be written explicitly as

$$\hat{\theta}_2 = R_N^{-1}\left[\frac{1}{N}\sum_{t=1}^{N}\frac{1}{C^2(q^{-1})}\phi(t - nc)y(t)\right], \tag{48}$$

where

$$R_N = \frac{1}{N}\sum_{t=1}^{N}\frac{1}{C^2(q^{-1})}\phi(t - nc)\phi^T(t). \tag{49}$$

It is straightforward to show that

$$\hat{\theta}_2 - \theta = R_N^{-1}\left[\frac{1}{N}\sum_{t=1}^{N}\frac{1}{C^2(q^{-1})}\phi(t - nc)v(t)\right], \tag{50}$$

and with

$$\hat{R}_N = \frac{1}{N}\sum_{t=1}^{N}\frac{1}{\hat{C}^2(q^{-1})}\phi(t - nc)\phi^T(t), \tag{51}$$

that

$$\begin{aligned}\hat{\hat{\theta}}_2 - \theta &= \hat{R}_N^{-1}\left[\frac{1}{N}\sum_{t=1}^{N}\frac{1}{\hat{C}^2(q^{-1})}\phi(t - nc)v(t)\right]\\ &= R_N^{-1}\left[\frac{1}{N}\sum_{t=1}^{N}\frac{1}{C^2(q^{-1})}\phi(t - nc)v(t)\right.\\ &\qquad - \left(\frac{2}{N}\sum_{t=1}^{N}\frac{1}{C^3(q^{-1})}[\phi(t - nc - 1), \ldots,\right.\\ &\qquad \left.\left.\phi(t - 2nc)]v(t)\right)(\hat{C} - C)\right] + 0(1/N)\\ &= \left(\hat{\theta}_2 - \theta\right) + 0(1/N),\end{aligned} \tag{52}$$

where

$$c = [c_1, \ldots, c_{nc}]^T, \quad \hat{c} = [\hat{c}_1, \ldots, \hat{c}_{nc}]^T.$$

Since $\hat{\theta}_2 - \theta = O(1/\sqrt{N})$, it follows from (52) that $\hat{\theta}_2$ and $\hat{\hat{\theta}}_2$ are asymptotically equivalent, which concludes the proof.

Note that here the choice of the number of MYW equations is not an issue, since optimality is achieved for m = na. However, the implementation of the OIV-2 estimator requires estimation of the $C(q^{-1})$ polynomial, while this can be avoided when implementing OIV-1. Estimating $C(q^{-1})$ is not an easy task and one often wants to avoid it, if possible. The relative advantages and disadvantages of the three estimators are discussed further in the following sections.

C. APPROXIMATE OIV-3

Let $\hat{C}(q^{-1})$ and $\hat{A}(q^{-1})$ denote consistent estimates of $C(q^{-1})$ and $A(q^{-1})$, respectively. Let $\hat{\hat{\theta}}_3$ be the approximate estimate obtained by replacing $C(q^{-1})$ and $A(q^{-1})$ by $\hat{C}(q^{-1})$ and $\hat{A}(q^{-1})$ in the OIV-3 estimate $\hat{\theta}_3$ [(6), (41)]. Then we can state the following theorem.

Theorem 7

Let assumptions A1-A3 hold true and assume that $\hat{c}_i - c_i = O(1/\sqrt{N})$, i = 1, ..., nc, and $\hat{a}_i - a_i = O(1/\sqrt{N})$, i = 1, ..., na. Then $\hat{\theta}_3$ and $\hat{\hat{\theta}}_3$ are asymptotically equivalent, provided that $m(N)/\sqrt{N} \to 0$ as $N \to \infty$.

Proof. Let

$$\hat{R}_N = \frac{1}{N} \sum_{t=1}^{N} \hat{z}_m(t)\phi^T(t), \tag{53}$$

where

$$\hat{z}_m(t) = \frac{\hat{A}(q^{-1})}{\hat{C}^2(q^{-1})}\begin{bmatrix} y(t - nc - 1) \\ \vdots \\ y(t - nc - m) \end{bmatrix}. \tag{54}$$

Then

$$\hat{R}_N = R_N + 0(1/\sqrt{N}), \qquad R_N = E\left\{z_m(t)\phi^T(t)\right\}, \tag{55}$$

where $z_m(t)$ is as defined in (5) with $G(q^{-1}) = A(q^{-1})/C^2(q^{-1})$. Thus,

$$\hat{\hat{\theta}}_3 \triangleq \left(\hat{R}_N^T\hat{R}_N\right)^{-1}\hat{R}_N^T \frac{1}{N}\sum_{t=1}^{N} \hat{z}_m(t)y(t)$$

$$= \theta + \left(\hat{R}_N^T\hat{R}_N\right)^{-1}\hat{R}_N^T \frac{1}{N}\sum_{t=1}^{N} \hat{z}_m(t)v(t). \tag{56}$$

Now

$$\left(\hat{R}_N^T\hat{R}_N\right)^{-1} = \left(R_N^TR_N + 0(m/\sqrt{N})\right)^{-1}$$

$$= \left(R_N^TR_N\right)^{-1} + 0(m/\sqrt{N}) \tag{57}$$

and

$$\frac{1}{N}\sum_{t=1}^{N} \hat{z}_m(t)v(t) = \frac{1}{N}\sum_{t=1}^{N} z_m(t)v(t) + 0(1/N). \tag{58}$$

Then

$$\hat{\hat{\theta}}_3 - \theta = \left[\left(R_N^TR_N\right)^{-1} + 0(m/\sqrt{N})\right]\left[R_N^T + 0(1/\sqrt{N})\right]$$

$$\times\left[\frac{1}{N}\sum_{t=1}^{N} z_m(t)v(t) + 0(1/N)\right]$$

$$= \left[\left(R_N^T R_N\right)^{-1} + 0(m/\sqrt{N})\right]\underbrace{\left[R_N^T \frac{1}{N}\sum_{t=1}^{N} z_m(t)v(t) + 0(m/\sqrt{N})\right]}_{0(m/\sqrt{N})}$$

$$= \left(\hat{\theta}_3 - \theta\right) + 0(m^2/N) \tag{59}$$

or

$$\hat{\hat{\theta}}_3 - \hat{\theta}_3 = 0(m^2/N). \tag{60}$$

Since $\left(\hat{\theta}_3 - \theta\right) = 0(m/\sqrt{N})$ [see (60)], we conclude that for $\hat{\hat{\theta}}_3$ and $\hat{\theta}_3$ to be asymptotically equivalent it is sufficient that

$$m/\sqrt{N} \to 0 \quad \text{as} \quad m,\ N \to \infty. \tag{61}$$

In this case $m^2/N = (m/\sqrt{N})^2$ goes to zero faster than the first term in (59), which concludes the proof.

The requirement in (61) that $m/\sqrt{N} \to 0$ is not restrictive. Note that $\hat{\theta}_3$ is not consistent if (61) is not satisfied [cf. (59)].

The behavior of $\tilde{P}_m$ (for OIV-3) as m increases is quite different from its behavior in the case of OIV-1. By specializing the results in Appendix B to the case of OIV-3, it can be readily shown that $\tilde{P}_m^{-1}$ obeys the discrete-time Lyapunov equation

$$\tilde{P}_{m+1}^{-1} - A\tilde{P}_m^{-1}A^T = bb^T, \tag{62}$$

where

$$A = \begin{bmatrix} -a_1 & \cdots & -a_{na} \\ 1 & & 0 \\ & \ddots & \\ 0 & & 1 \end{bmatrix},$$

$$b = \frac{1}{\lambda} E\left\{\phi(t)\frac{1}{C(q^{-1})} e(t - nc - 1)\right\}.$$

From the above equations we conclude that the convergence rate of P_m depends on the zeros of $A(q^{-1})$, but not on the zeros of $C(q^{-1})$. The reverse is true for OIV-1; see the discussion in Section V. More specifically, $|\tilde{P}_\infty - \tilde{P}_m| \sim |\lambda_{max}|^{2m}$, where λ_{max} is the zero of $A(q^{-1})$ with the largest modulus. Thus, when $A(q^{-1})$ has roots closer to the unit circle than the zeros of $C(q^{-1})$, we expect $\tilde{P}_m$ to converge faster for OIV-1 than for OIV-3 [and vice versa when the zeros of $C(q^{-1})$ are closer to the unit circle than the zeros of $A(q^{-1})$].

VIII. IMPLEMENTATION OF THE OPTIMAL IV MULTISTEP ESTIMATORS

A. *THE OIV-1 ALGORITHM*

Let us denote by $r_v(\tau)$ and $R_z(\tau)$ the covariances of $v(t)$ and $z_m(t)$, respectively:

$$r_v(\tau) = E\{v(t)v(t - \tau)\}, \tag{63}$$

$$R_z(\tau) = E\left\{z_m(t)z_m^T(t - \tau)\right\}. \tag{64}$$

Next note that

$$\begin{aligned} S_m &= E\left\{C(q^{-1})z_m(t)C(q^{-1})z_m^T(t)\right\} \\ &= E\left\{\sum_{i=0}^{nc}\sum_{j=0}^{nc} c_i c_j z_m(t - i)z_m^T(t - j)\right\} \\ &= \sum_{i=0}^{nc}\sum_{j=0}^{nc} c_i c_j R_z(j - i) \\ &= \sum_{\tau=-nc}^{nc}\left(\sum_{j=0}^{nc} c_j c_{j+\tau}\right)R_z(\tau) \\ &= \frac{1}{\lambda^2}\sum_{\tau=-nc}^{nc} r_v(\tau)R_z(\tau). \end{aligned} \tag{65}$$

In the following we will omit the factor $1/\lambda^2$ appearing in (65) since the IV estimates in (6) are invariant to scaling of the weighting matrix Q. From (65) it follows that we can consistently estimate the optimal weighting matrix by $\hat{S}_m^{-1}$, where

$$\hat{S}_m = \sum_{\tau=-nc}^{nc} \hat{r}_v(\tau)\hat{R}_z(\tau), \tag{66}$$

and where $\hat{r}_v(\tau)$ and $\hat{R}_v(\tau)$ are the sample covariances:

$$\hat{r}_v(\tau) = \frac{1}{N}\sum_{t=\tau}^{N} v(t)v(t-\tau) = \hat{r}_v(-\tau), \tag{67}$$

$$\hat{R}_z(\tau) = \frac{1}{N}\sum_{t=\tau}^{N} z_m(t)z_m^T(t-\tau) = \hat{R}_z^T(-\tau). \tag{68}$$

Note that (66) provides a method for estimating S without explicit estimation of the $\{c_i\}$ parameters. However, to estimate $\hat{r}_v(\tau)$ via (67) we need to compute (an estimate of)

$$v(t) = A(q^{-1})y(t). \tag{69}$$

An alternative way of computing $\hat{r}_v(\tau)$ follows from (69):

$$\hat{r}_v(\tau) = \sum_{i=0}^{na}\sum_{j=0}^{na} \hat{a}_i\hat{a}_j\hat{r}_y(\tau+i-j). \tag{70}$$

Note that (67), (69), and (70) require knowledge of the $\{a_i\}$ parameters. Since $A(q^{-1})$ is not known a priori, we must use a multistep procedure. We first estimate $\{a_i\}$ using (6) with $Q = I$ [and $G(q^{-1}) = 1$]. The result will be problem dependent, but generally m will be considerably larger than na (see [9]). This gives a consistent, although not efficient, estimate of the AR parameters. These estimates can now be used to compute $\hat{r}_v(\tau)$ via (67) and (69) or (70). Next we compute $\hat{S}_m$ [(66)]

and use it in the OIV-1 procedure to obtain the final (asymptotically efficient) estimate of the AR parameters.

We can now summarize the proposed implementation of OIV-1 estimator.

(i) Estimate $\{a_i\}$ by (6) with $Q = I$, $G(q^{-1}) = 1$, and $m \geq na$.

(ii) Compute $\hat{R}_z(\tau)$ by (67) and $\hat{r}_v(\tau)$ by (69) or (70) using the $\{a_i\}$ estimates from step (i), and then compute $\hat{S}_m$ (66).

(iii) Compute the square root $\hat{S}_m^{-1/2}$ of $\hat{S}_m^{-1}$; then solve (6) with $Q^{1/2} = \hat{S}_m^{-1/2}$ to obtain the final $\{a_i\}$ estimates.

Note that the computation of $\hat{S}_m$ via (66) does not guarantee that $\hat{S}_m$ will be a positive-definite matrix. It may happen, therefore, that $\hat{S}_m^{-1/2}$ does not exist. This is unlikely to occur for large N, but is quite likely for small sample sizes [especially if C(z) has roots close to the unit circle].

The following is a procedure for handling the case where $\hat{S}_m$ is not positive definite. Let $\{\lambda_i\}_{i=1}^m$ be the ordered eigenvalues of $\hat{S}_m$, $\lambda_1 \geq \lambda_2 \geq \cdots \geq \lambda_m$, and let $\{v_i\}_{i=1}^m$ be the corresponding eigenvectors. Let

$$\begin{aligned} \lambda_k &\geq \epsilon, \quad k = 1, \ldots, n, \\ \lambda_k &< \epsilon, \quad k = n+1, \ldots, m, \end{aligned} \tag{71}$$

with ϵ being a (small) positive number. Further, let $\mathscr{C}$ be the class of positive-definite matrices with eigenvalues larger than or equal to ϵ. Then, according to Lemma D.1, the Euclidean distance between $\hat{S}_m$ and the elements of $\mathscr{C}$ is minimal for the matrix $\tilde{S}_m$ given by

$$\tilde{S}_m = V \cdot [\mathrm{diag}(\lambda_1, \ldots, \lambda_n, \epsilon, \ldots, \epsilon)] \cdot V^T, \tag{72}$$

where $V = [v_1, \ldots, v_m]$. We will use

$$\tilde{S}_m^{-1/2} = \operatorname{diag}\left(\frac{1}{\sqrt{\lambda_1}}, \ldots, \frac{1}{\sqrt{\lambda_n}}, \frac{1}{\sqrt{\epsilon}}, \ldots, \frac{1}{\sqrt{\epsilon}}\right)V^T \tag{73}$$

in (6) instead of $\hat{S}_m^{-1/2}$, which may not exist. Since $\tilde{S}_m$ must be a consistent estimate of $\lambda^2 S_m$, ϵ must go to zero as N tends to infinity. To guarantee consistency we may set $\epsilon = 1/N^\beta$, $\beta > 0$. As $N \to \infty$ we will have $\tilde{S}_m \to \hat{S}_m$, where $\hat{S}_m$ is a consistent estimate of $\lambda^2 S_m$. Concerning the choice of β, we note that the smaller ϵ, the smaller is the distance between $\tilde{S}_m$ and $\hat{S}_m$ (cf. Lemma D.1). However, too small an ϵ may lead to ill-conditioning problems. Thus, ϵ should be chosen as a compromise between accuracy of the solution and numerical stability. Finally, note that if the estimated covariance matrix $\hat{S}_m$ happens to have negative eigenvalues, then we may suspect that the $\{a_i\}$ estimates obtained in step (i) were poor. We may then wish to repeat steps (ii)-(iii) using in step (ii) the improved estimates of step (iii).

B. *THE OIV-2 ALGORITHM*

The computation of the OIV-2 estimates requires the estimation of the $\{c_i\}$ parameters. There are, of course, many different ways in which this could be done. We consider here one such method based on factorization of the MA spectrum [34-36].

Let $S_v(z)$ denote the spectral density function of v(t) [(2), (69)]. We have

$$S_v(z) \triangleq \sum_{k=-nc}^{nc} r_v(k) z^{-k} = \lambda^2 C(z) C(z^{-1}), \tag{74}$$

where $r_v(k)$ denotes the covariance of v(t) at lag k (63). In other words, the C(z) polynomial is the spectral factor of the spectrum of $v(t)$. This suggests the following procedure for estimating the $\{c_i\}$ parameters.

(a) Estimate the $\{a_i\}$ parameters using (6) with $Q = I$, $G(q^{-1}) = 1$, and $m \geq na$.

(b) Compute the sample covariances $\hat{r}_v(k)$, $k = 0, \ldots, nc$, using (67) and (69) or (70).

(c) Perform spectral factorization of $\hat{S}_v(z) = \Sigma_{k=-nc}^{nc} \hat{r}_v(k) z^{-k}$ to obtain $\left\{\hat{c}_i\right\}$.

Note that the sample covariance sequence $\left\{\hat{r}_v(0), \ldots, \hat{r}_v(nc), 0, 0, \ldots\right\}$ is not guaranteed to be positive definite. Thus, $\hat{S}_v(z)$ may not be factorable. This may happen in the small sample case, especially when $C(z)$ has roots close to the unit circle. However, note that OIV-2 requires an estimate of $C^2(q^{-1})$ ra rather than of $C(q^{-1})$. We can always obtain a consistent estimate of $C^2(q^{-1})$ by factoring $\hat{S}_v^2(z)$, since

$$\hat{S}_v^2(e^{j\omega}) = \hat{C}^2(e^{j\omega})\hat{C}^2(e^{-j\omega}) \geq 0, \quad \text{for all } \omega, \tag{75}$$

even though $\hat{S}_v(e^{j\omega})$ may be negative for some values of ω.

We can now summarize the proposed implementation of OIV-2.

(i) Estimate $C^2(q^{-1})$ using the spectral factorization method described above. Let $\hat{G}(q^{-1}) = 1/\hat{C}^2(q^{-1})$.

(ii) Estimate the AR parameters using (6) with $Q = I$, $m = na$, and $\hat{G}(q^{-1})$ from step (i).

Note that it is possible to iterate this procedure by using the AR parameters obtained in step (ii) to improve the estimate of $C^2(q^{-1})$, by repeating step (i) (the factorization method) with the new $\left\{\hat{a}_i\right\}$ parameters.

C. THE OIV-3 ALGORITHM

The computation of the OIV-3 estimates is very similar to that of OIV-2. The only difference is that $\hat{G}(q^{-1}) = \hat{A}(q^{-1})/\hat{C}^2(q^{-1})$, where $\hat{A}(q^{-1})$ is the current estimate of $A(q^{-1})$ [obtained from step (a) in the first iteration of the algorithm, or from the previous step (ii) in the case of reiteration]. Recall also that in the case of OIV-3 m should be (much) larger than na if optimal accuracy is desired.

D. COMPUTATIONAL REQUIREMENTS

The following is a brief summary of the number of arithmetic operations (i.e., multiplications and additions) required by each of the algorithms described above.

OIV-1

Step (i) requires approximately $(m + na)N$ operations to compute the sample covariances and approximately na^2m operations to solve for the initial estimate. Step (ii) requires approximately $(na)^2nc$ operations to evaluate $\hat{r}_v(\tau)$ using (70) or approximately $(na + nc)N$ operations using (67), (69). The computation of $\hat{S}_m$ requires approximately ncm^2 operations. Step (iii) requires approximately $3m^3$ operations. A recursive QR algorithm which appears to be useful for solving (6) is presented in Appendix E.

OIV-2

Step (a) requires approximately $(m + na)N + m^3$ operations, as in the case of OIV-1. Step (b) involves the computation of $\hat{r}_v(\tau)$, which requires either approximately $(na)^2nc$ or approximately $(na + nc)N$ operations. For step (c) the computational requirements will depend on the particular factorization technique. Step (ii) requires approximately $2(nc + na)N + (na)^3$ operations.

OIV-3

Steps (a)-(c) are same as for OIV-2. Step (ii) requires approximately $(2nc + na)N$ operations to perform the filtering and approximately $(na + m)N$ operations to compute the necessary covariances. Finally, solving the Yule-Walker equations requires approximately na^2m operations.

In summary:

OIV-1: $(m + na)N + 3m^3 + ncm^2 + na^2m$

[or $(m + 2na + nc)N + 3m^3 + ncm^2 + na^2m$]

OIV-2: $(m + 2na + 2nc)N + na^2m + na^2(nc + na)$

[or $(m + 4na + 4nc)N + na^2m + na^3$]

OIV-3: $(2m + 3na + 2nc)N + na^2(nc + m)$

[or $(3m + 5na + 3nc)N + na^2m$]

Note also that reiteration does not require much computation, since the sample covariances need to be computed only once. Iteration of OIV-2 and OIV-3 is more costly since the data need to be refiltered and some sample covariances recomputed at each iteration.

IX. NUMERICAL EXAMPLES

In this section we present some selected results of computer simulations that illustrate the behavior of the OIV algorithms discussed earlier. Tables 4-11 summarize results based on 100 independent Monte Carlo runs performed for each of the test cases described below. Each of the tables contains the means and standard deviations (as well as the square roots of the mean-squared errors) of the AR parameter estimates obtained by applying the MYW method, OIV-1, OIV-2, and OIV-3 algorithms to simulated data. The OIV algorithms were used with different values of m and iterated three times. The theoretical covariances for the case m = 40 are missing from the table since their evaluation was difficult due to some features of the particular program we were using.

Note that OIV-2 was run for values of m different from m = na. The symptotic theory shows that OIV-2 is optimal only for m = na, not for m > na. (Indeed, for m > na the optimal matrix is $Q = S_m^{-1}$, while OIV-2 uses Q = I; recall that for m = na Q is irrelevant.) However, in the finite-data case we found that increasing m tended to make the algorithm more robust. For m = na the estimates provided by the MYW method are very inaccurate; thus, the OIV-2 algorithm is poorly initialized. Iteration will eventually improve the accuracy of the estimates.

In the first two cases the data were the sum of a second-order autoregressive process and white noise:

$$y(t) = x(t) + n(t), \tag{76}$$

where

$$x(t) = -a_z x(t - 1) - a_2 x(t - 2) + w(t), \tag{77}$$

and where w(t) and n(t) are mutually uncorrelated white noise processes whose variances were chosen to give the desired signal-to-noise ratio (SNR = Var{x(t)}/Var{n(t)}). As is well known, y(t) has an equivalent ARMA (2, 2) representation. The zeros of the MA part can be shown to be farther away from the unit circle than the zeros of the AR part. As the SNR decreases, the MA zeros approach the AR zeros.

Case 1: Narrowband, High SNR

$$A(z) = 1 - 1.4z + 0.95z^2 \quad (\text{zeros at } 0.975e^{\pm j44.1°}),$$
$$\text{SNR} = 20 \text{ dB}, \quad N = 4096. \tag{78}$$

The MA polynomial of the equivalent ARMA representation of this process is

$$C(z) = 1 - 0.3155z + 0.1233z^2 \quad (\text{zeros at } 0.351e^{\pm j63.3°}). \tag{79}$$

The results are summarized in Tables 4 and 5. In this high-SNR case the experimental results are very close to the asymptotic bounds. Note that OIV-1 converges quickly [since C(z) has zeros close to the origin]. The accuracy of the MYW method is comparable in this case to that of the optimal IV methods. This implies that reiteration is not needed, since the initial estimates are good. Note also the slow convergence of P_{OIV-1} to $P_{CR} = P_{OIV-2}$ as m increases. This slow convergence is to be expected since A(z) has zeros close to the unit circle.

Case 2: Narrowband, Low SNR

A(z) and N are as in Case 1, but now SNR = 0 dB, leading to

$$C(z) = 1 - 1.20955z + 0.726837z^2 \quad (\text{zeros at } 0.853e^{\pm j44.8°}). \tag{80}$$

The results are summarized in Tables 6 and 7.

In this low-SNR case the experimental results are quite different from the theoretical bounds. Note that in this case the OIV methods are quite a bit better than the MYW algorithm. Note also that iterating the OIV algorithms tends to reduce the bias, although not necessarily the variance, of the estimates.

The following two test cases involve ARMA rather than AR-plus-noise processes.

Case 3

$$\frac{C(z)}{A(z)} = \frac{1 - 1.8z + 0.95z^2}{1 - 1.5z + 0.7z^2}, \qquad N = 4096. \tag{81}$$

In this case the zeros of the MA part ($0.975e^{\pm j22.5°}$) are closer to the unit circle than the zeros of the AR part ($0.837e^{\pm j26.3°}$). The results are summarized in Tables 8 and 9. Note that in this case too the OIV algorithms behave considerably better than the MYW method.

Case 4

$$\frac{C(z)}{A(z)} = \frac{1 - 0.9747z + 0.95z^2}{1 - 1.688z + 0.95z^2}, \qquad N = 4096. \tag{82}$$

In this case the zeros of the MA part ($0.975e^{\pm j60°}$) and the AR part ($0.975e^{\pm j30°}$) are at the same distance from the unit circle, but their phase angles are well separated. The results are summarized in Tables 10 and 11.

Finally, we present some comments about the practical experience with these algorithms.

Equation (6) was solved by an algorithm based on the MINFIT subroutine in EISPACK, which computes the singular-value decomposition using Householder bidiagonalization and an AR algorithm. Solutions corresponding to singular values less than 10^{-8} times the largest singular value were set to zero (in effect decreasing the assumed rank of the MYW equations and producing the minimum-norm solution of the undetermined set of equations).

OIV-1: The algorithm was implemented as described in Section IV and appears to be quite robust. Since our simulations involved relatively long data records, we did not encounter problems with $\hat{S}_m$ being non-positive definite. Thus, we did not

Table 4. *Experimental and Theoretical Estimation Accuracy for Case 1, Parameter a_1*

	m			
	2	4	10	40
MYW				
Mean ± std. dev.	-1.40 ± 0.00531	-1.40 ± 0.00530	-1.40 ± 0.00530	-1.40 ± 0.00611
mse = $[\text{variance} + (\text{bias})^2]^{1/2}$	0.00542	0.00541	0.00537	0.00619
Theoretical std. dev.	0.00536	0.00533	0.00550	--
OIV-1				
Iteration 1				
Mean ± std. dev.	--	-1.40 ± 0.00538	-1.40 ± 0.00529	-1.40 ± 0.00537
mse	--	0.00549	0.00539	0.00548
Iteration 2				
Mean ± std. dev.	--	-1.40 ± 0.00545	-1.40 ± 0.00535	-1.40 ± 0.00534
mse	--	0.00556	0.00546	0.00545
Iteration 3				
Mean ± std. dev.	--	-1.40 ± 0.00544	-1.40 ± 0.00541	-1.40 ± 0.00531
mse	--	0.00555	0.00552	0.00542
Theoretical std. dev.	0.00536	0.00532	0.00531	--

OIV-2				
Iteration 1				
Mean ± std. dev.	*-1.40 ± 0.00540*	*-1.40 ± 0.00547*	*-1.40 ± 0.00552*	*-1.40 ± 0.00631*
mse	*0.00550*	*0.00558*	*0.00559*	*0.00638*
Iteration 2				
Mean ± std. dev.	*-1.40 ± 0.00547*	*-1.40 ± 0.00538*	*-1.40 ± 0.00544*	*-1.40 ± 0.00632*
mse	*0.00557*	*0.00550*	*0.00551*	*0.00639*
Iteration 3				
Mean ± std. dev.	*-1.40 ± 0.00530*	*-1.40 ± 0.00540*	*-1.40 ± 0.00548*	*-1.40 ± 0.00632*
mse	*0.00541*	*0.00551*	*0.00555*	*0.00639*
Theoretical std. dev.	*0.00532*	--	--	--
OIV-3				
Iteration 1				
Mean ± std. dev.	--	*-1.40 ± 0.0135*	*-1.40 ± 0.0104*	*-1.40 ± 0.00607*
mse	--	*0.0139*	*0.0106*	*0.00630*
Iteration 2				
Mean ± std. dev.	--	*-1.40 ± 0.0132*	*-1.40 ± 0.0103*	*-1.40 ± 0.00623*
mse	--	*0.0133*	*0.0103*	*0.00633*
Iteration 3				
Mean ± std. dev.	--	*-1.40 ± 0.0140*	*-1.40 ± 0.0106*	*-1.40 ± 0.00618*
mse		*0.0141*	*0.0106*	*0.00629*
Theoretical std. dev.	*0.0275*	*0.0122*	*0.0087*	--

Table 5. *Experimental and Theoretical Estimation Accuracy for Case 1, Parameter* a_2

	m			
	2	4	10	40
MYW				
Mean ± std. dev.	0.949 ± 0.00476	0.9491 ± 0.004792	0.9492 ± 0.004974	0.9492 ± 0.005823
mse = $[\text{variance} + (\text{bias})^2]^{1/2}$	0.004863	0.004882	0.005031	0.005878
Theoretical std. dev.	0.00507	0.00508	0.00534	--
OIV-1				
Iteration 1				
Mean ± std. dev.	--	0.9490 ± 0.004816	0.9490 ± 0.004784	0.9490 ± 0.004792
mse	--	0.004916	0.004883	0.004894
Iteration 2				
Mean ± std. dev.	--	0.9490 ± 0.004856	0.9490 ± 0.004800	0.9490 ± 0.004800
mse	--	0.004955	0.004800	0.004902
Iteration 3				
Mean ± std. dev.	--	0.9490 ± 0.004840	0.9490 ± 0.004800	0.9490 ± 0.004816
mse	--	0.004939	0.004899	0.004918
Theoretical std. dev.	0.00507	0.00506	0.00506	--

OIV-2				
Iteration 1				
Mean ± std. dev.	0.949 ± 0.004981	0.9490 ± 0.005058	0.9493 ± 0.005164	0.9493 ± 0.006057
mse	0.005096	0.005154	0.005217	0.006101
Iteration 2				
Mean ± std. dev.	0.949 ± 0.004966	0.9490 ± 0.004989	0.9493 ± 0.005134	0.9493 ± 0.006082
mse	0.005080	0.005086	0.005188	0.006126
Iteration 3				
Mean ± std. dev.	0.949 ± 0.004958	0.9490 ± 0.005012	0.9493 ± 0.005134	0.9493 ± 0.006044
mse	0.005073	0.005109	0.005188	0.006089
Theoretical std. dev.	0.00506	--	--	--
OIV-3				
Iteration 1				
Mean ± std. dev.	--	0.9467 ± 0.01227	0.9480 ± 0.007988	0.9480 ± 0.005559
mse	--	0.01270	0.008243	0.005912
Iteration 2				
Mean ± std. dev.	--	0.9493 ± 0.01229	0.9494 ± 0.008136	0.9487 ± 0.005587
mse	--	0.01231	0.008155	0.005739
Iteration 3				
Mean ± std. dev.	--	0.9493 ± 0.01285	0.9495 ± 0.008342	0.948 ± 0.005614
mse		0.01286	0.008360	0.005771
Theoretical std. dev.	0.01394	0.0117	0.00767	--

Table 6. Experimental and Theoretical Estimation Accuracy for Case 2, Parameter a_1

	m			
	2	4	10	40
MYW				
Mean ± std. dev.	1.729 ± 2.597	−0.926 ± 0.390	−0.811 ± 0.275	−0.5521 ± 0.1883
mse = $[\text{variance} + (\text{bias})^2]^{1/2}$	2.618	0.614	0.0650	0.8686
Theoretical std. dev.	0.05406	0.0130	0.0092	--
OIV-1				
Iteration 1				
Mean ± std. dev.	--	−1.16 ± 0.399	−1.16 ± 0.307	−0.9154 ± 0.2834
mse	--	0.468	0.388	0.5614
Iteration 2				
Mean ± std. dev.	--	−1.21 ± 0.397	−1.26 ± 0.301	−1.152 ± 0.2800
mse	--	0.442	0.333	0.3740
Iteration 3				
Mean ± std. dev.	--	−1.22 ± 0.391	−1.28 ± 0.288	−1.266 ± 0.2372
mse	--	0.429	0.311	0.2722
Theoretical std. dev.	0.05406	0.0127	0.00786	--

OIV-2				
Iteration 1				
Mean ± std. dev.	-0.8098 ± 2.269	-1.27 ± 0.335	-1.29 ± 0.270	-1.192 ± 0.2370
mse	2.344	0.356	0.292	0.3154
Iteration 2				
Mean ± std. dev.	-1.096 ± 0.9459	-1.30 ± 0.317	-1.34 ± 0.275	-1.375 ± 0.1276
mse	0.9935	0.334	0.281	0.1301
Iteration 3				
Mean ± std. dev.	-0.9804 ± 1.240	-1.32 ± 0.315	-1.34 ± 0.346	-1.405 ± 0.1195
mse	1.309	0.327	0.351	0.1197
Theoretical std. dev.	0.00674	--	--	--
OIV-3				
Iteration 1				
Mean ± std. dev.	--	-1.15 ± 0.423	-1.19 ± 0.296	-0.9724 ± 0.2806
mse	--	0.493	0.362	0.5115
Iteration 2				
Mean ± std. dev.	--	-1.21 ± 0.371	-1.291 ± 0.2847	-1.198 ± 0.2507
mse	--	0.417	0.3047	0.3222
Iteration 3				
Mean ± std. dev.	--	-1.27 ± 0.328	-1.313 ± 0.2953	-1.281 ± 0.2029
mse		0.354	0.3079	0.2351
Theoretical std. dev.	0.01684	0.0156	0.01011	--

Table 7. Experimental and Theoretical Estimation Accuracy for Case 2, Parameter a_2

	m			
	2	4	10	40
MYW				
Mean ± std. dev.	0.9279 ± 0.672	0.5293 ± 0.3846	0.4763 ± 0.2238	0.2484 ± 0.1413
mse = $[\text{variance} + (\text{bias})^2]^{1/2}$	2.672	0.5700	0.5239	0.7157
Theoretical std. dev.	0.04070	0.0211	0.00945	--
OIV-1				
Iteration 1				
Mean ± std. dev.	--	0.6980 ± 0.4347	0.7543 ± 0.2424	0.5358 ± 0.2247
mse	--	0.5025	0.3115	0.4712
Iteration 2				
Mean ± std. dev.	--	0.7405 ± 0.4275	0.8369 ± 0.2354	0.7375 ± 0.2229
mse	--	0.4761	0.2399	0.3079
Iteration 3				
Mean ± std. dev.	--	0.7541 ± 0.4232	0.8563 ± 0.2209	0.8342 ± 0.1828
mse	--	0.4664	0.2399	0.2164
Theoretical std. dev.	0.04070	0.0194	0.00778	--

OIV-2				
Iteration 1				
Mean ± std. dev.	0.7778 ± 0.6499	0.8279 ± 0.3295	0.8676 ± 0.2074	0.7636 ± 0.1945
mse	0.6723	0.3515	0.2231	0.2694
Iteration 2				
Mean ± std. dev.	0.8574 ± 0.7826	0.8695 ± 0.2966	0.9271 ± 0.1424	0.9283 ± 0.08034
mse	0.7880	0.3074	0.1442	0.08321
Iteration 3				
Mean ± std. dev.	0.8611 ± 0.4633	0.8934 ± 0.2389	0.9440 ± 0.07594	0.9504 ± 0.03850
mse	0.4717	0.2455	0.07618	0.03850
Theoretical std. dev.	0.0067	--	--	--
OIV-3				
Iteration 1				
Mean ± std. dev.	--	0.7661 ± 0.2853	0.7815 ± 0.2326	0.5844 ± 0.2260
mse	--	0.3395	0.2872	0.4299
Iteration 2				
Mean ± std. dev.	--	0.8008 ± 0.3085	0.8603 ± 0.2122	0.7783 ± 0.2025
mse	--	0.3426	0.2304	0.2655
Iteration 3				
Mean ± std. dev.	--	0.8337 ± 0.2615	0.8904 ± 0.2102	0.8443 ± 0.1508
mse	--	0.2862	0.2185	0.1841
Theoretical std. dev.	0.03147	0.0157	0.01064	--

Table 8. *Experimental and Theoretical Estimation Accuracy for Case 3, Parameter* a_1

	m			
	2	4	10	40
MYW				
Mean ± std. dev.	0.3401 ± 11.98	-0.9641 ± 0.4967	-0.8148 ± 0.2602	-0.3607 ± 0.1214
mse = $[\text{variance} + (\text{bias})^2]^{1/2}$	12.12	0.7303	0.7329	1.146
Theoretical std. dev.	2.599	0.547	0.2045	--
OIV-1				
Iteration 1				
Mean ± std. dev.	--	-1.285 ± 0.6152	-1.452 ± 0.06578	-1.302 ± 0.09564
mse	--	0.6517	0.08153	0.2202
Iteration 2				
Mean ± std. dev.	--	-1.320 ± 0.6188	-1.491 ± 0.04059	-1.478 ± 0.03514
mse	--	0.6443	0.04164	0.04164
Iteration 3				
Mean ± std. dev.	--	-1.333 ± 0.6047	-1.491 ± 0.04180	-1.483 ± 0.03768
mse	--	0.6272	0.04276	0.04137
Theoretical std. dev.	2.599	0.491	0.03445	--

OIV-2				
Iteration 1				
Mean ± std. dev.	−0.9022 ± 7.139	−1.454 ± 0.2030	−1.484 ± 0.04267	−1.356 ± 0.1225
mse	7.164	0.2082	0.04562	0.1891
Iteration 2				
Mean ± std. dev.	−1.569 ± 1.134	−1.475 ± 0.2055	−1.496 ± 0.05307	−1.505 ± 0.05697
mse	1.136	0.2070	0.05325	0.05724
Iteration 3				
Mean ± std. dev.	−1.492 ± 0.1007	−1.474 ± 0.2228	−1.506 ± 0.05888	−1.514 ± 0.06048
mse	0.1010	0.2243	0.05916	0.06199
Theoretical std. dev.	0.0139	--	--	--
OIV-3				
Iteration 1				
Mean ± std. dev.	--	−1.429 ± 0.2372	−1.445 ± 0.07617	−1.152 ± 0.1534
mse	--	0.2476	0.09402	0.3800
Iteration 2				
Mean ± std. dev.	--	−1.459 ± 0.2476	−1.466 ± 0.06375	−1.430 ± 0.1044
mse		0.2510	0.07213	0.1258
Iteration 3				
Mean ± std. dev.	--	−1.467 ± 0.2483	−1.466 ± 0.09093	−1.423 ± 0.089716
mse	--	0.2505	0.09710	0.1166
Theoretical std. dev.	0.01629	0.0160	0.01405	--

Table 9. Experimental and Theoretical Estimation Accuracy for Case 3, Parameter a_2

	m			
	2	4	10	40
MYW				
Mean ± std. dev.	-0.5767 ± 9.21	0.3323 ± 0.3235	0.2114 ± 0.1877	-0.07178 ± 0.1006
mse = $[\text{variance} + (\text{bias})^2]^{1/2}$	9.229	0.4897	0.5234	0.7783
Theoretical std. dev.	1.95843	0.335	0.10016	--
OIV-1				
Iteration 1				
Mean ± std. dev.	--	0.5733 ± 0.3739	0.6711 ± 0.04926	0.5656 ± 0.07516
mse	--	0.3948	0.05712	0.1540
Iteration 2				
Mean ± std. dev.	--	0.5961 ± 0.3756	0.6953 ± 0.03173	0.6889 ± 0.02725
mse	--	0.3897	0.03129	0.02942
Iteration 3				
Mean ± std. dev.	--	0.6051 ± 0.3654	0.6956 ± 0.03173	0.6956 ± 0.02699
mse	--	0.3775	0.03203	0.02734
Theoretical std. dev.	1.95843	0.301	0.02911	--

OIV-2				
Iteration 1				
Mean ± std. dev.	*-0.7038 ± 13.22*	*0.6674 ± 0.1260*	*0.6835 ± 0.03746*	*0.5894 ± 0.09895*
mse	*13.29*	*0.1302*	*0.04093*	*0.1484*
Iteration 2				
Mean ± std. dev.	*0.7563 ± 0.6148*	*0.6845 ± 0.1222*	*0.7003 ± 0.04660*	*0.7137 ± 0.04116*
mse	*0.6173*	*0.1232*	*0.04660*	*0.04337*
Iteration 3				
Mean ± std. dev.	*0.6952 ± 0.08419*	*0.6912 ± 0.1350*	*0.7095 ± 0.05817*	*0.7373 ± 0.06200*
mse	*0.08433*	*0.1352*	*0.05893*	*0.07234*
Theoretical std. dev.	*0.0138*	--	--	--
OIV-3				
Iteration 1				
Mean ± std. dev.	--	*0.6626 ± 0.1356*	*0.6610 ± 0.05672*	*0.4685 ± 0.1128*
mse	--	*0.1407*	*0.06885*	*0.2575*
Iteration 2				
Mean ± std. dev.	--	*0.6771 ± 0.1479*	*0.6815 ± 0.04338*	*0.6543 ± 0.06464*
mse	--	*0.1497*	*0.04718*	*0.07915*
Iteration 3				
Mean	--	*0.6914 ± 0.1523*	*0.6819 ± 0.06095*	*0.6598 ± 0.05870*
mse	--	*0.1526*	*0.06359*	*0.07113*
Theoretical std. dev.	*0.03035*	*0.0185*	*0.01408*	--

Table 10. Experimental and Theoretical Estimation Accuracy for Case 4, Parameter a_1

	m			
	2	4	10	40
MYW				
Mean ± std. dev.	−1.687 ± 0.007969	−1.687 ± 0.005607	−1.687 ± 0.005902	−1.687 ± 0.006709
mse = $[\text{variance} + (\text{bias})^2]^{1/2}$	0.008070	0.005721	0.005983	0.006749
Theoretical std. dev.	0.00788	0.00585	0.00581	--
OIV-1				
Iteration 1				
Mean ± std. dev.	--	−1.687 ± 0.005716	−1.687 ± 0.005296	−1.686 ± 0.009568
mse	--	0.00582	0.005359	0.009733
Iteration 2				
Mean ± std. dev.	--	−1.687 ± 0.005716	−1.687 ± 0.005325	−1.688 ± 0.01345
mse	--	0.005822	0.005387	0.01345
Iteration 3				
Mean ± std. dev.	--	−1.687 ± 0.005716	−1.687 ± 0.005325	−1.689 ± 0.01890
mse	--	0.005822	0.005387	0.01890
Theoretical std. dev.	0.00788	0.00573	0.0052	--

OIV-2				
Iteration 1				
Mean ± std. dev.	-1.688 ± 0.005354	-1.688 ± 0.005325	-1.688 ± 0.005607	-1.688 ± 0.006522
mse	0.005397	0.005371	0.005651	0.006550
Iteration 2				
Mean ± std. dev.	-1.687 ± 0.005552	-1.688 ± 0.005468	-1.688 ± 0.00524	-1.688 ± 0.006709
mse	0.005601	0.005512	0.005566	0.006736
Iteration 3				
Mean ± std. dev.	-1.688 ± 0.005383	0.005677	-1.688 ± 0.005552	-1.688 ± 0.006663
mse	0.005423	0.005677	0.005594	0.006690
Theoretical std. dev.	0.00156	--	--	--
OIV-3				
Iteration 1				
Mean ± std. dev.	--	-1.687 ± 0.01697	-1.687 ± 0.009324	-1.686 ± 0.006355
mse	--	0.01701	0.009392	0.006620
Iteration 2				
Mean ± std. dev.	--	-1.692 ± 0.01869	-1.689 ± 0.009072	-1.687 ± 0.005980
mse		0.01901	0.009116	0.006050
Iteration 3				
Mean ± std. dev.	--	-1.692 ± 0.02019	-1.689 ± 0.009854	-1.687 ± 0.006132
mse	--	0.02046	0.009880	0.006221
Theoretical std. dev.	0.01771	0.01291	0.00794	--

Table 11. Experimental and Theoretical Estimation Accuracy for Case 4, Parameter a_2

	m			
	2	4	10	40
MYW				
Mean ± std. dev.	0.9488 ± 0.007186	0.9487 ± 0.006604	0.9488 ± 0.006789	0.9591 ± 0.007339
mse = $[\text{variance} + (\text{bias})^2]^{1/2}$	0.007325	0.006738	0.006885	0.007399
Theoretical std. dev.	0.00606	0.00561	0.00556	--
OIV-1				
Iteration 1				
Mean ± std. dev.	--	0.9487 ± 0.006703	0.9488 ± 0.006391	0.9467 ± 0.01466
mse	--	0.006836	0.006499	0.01503
Iteration 2				
Mean ± std. dev.	--	0.9487 ± 0.006691	0.9488 ± 0.006373	0.9497 ± 0.01053
mse	--	0.006825	0.006480	0.01053
Iteration 3				
Mean ± std. dev.	--	0.9487 ± 0.006674	0.9488 ± 0.006336	0.9481 ± 0.01965
mse	--	0.006808	0.006445	0.01974
Theoretical std. dev.	0.00606	0.00552	0.00516	--

OIV-2				
Iteration 1				
Mean ± std. dev.	0.9489 ± 0.006385	0.9489 ± 0.006409	0.9491 ± 0.006616	0.9492 ± 0.007355
mse	0.006478	0.006498	0.006684	0.007395
Iteration 2				
Mean ± std. dev.	0.9489 ± 0.006445	0.9489 ± 0.006415	0.9491 ± 0.006593	0.9492 ± 0.007387
mse	0.006544	0.00650	0.006658	0.007426
Iteration 3				
Mean ± std. dev.	0.9489 ± 0.006397	0.9489 ± 0.006397	0.9491 ± 0.006622	0.9492 ± 0.007387
mse	0.006500	0.006483	0.006687	0.007426
Theoretical std. dev.	0.00156	--	--	--
OIV-3				
Iteration 1				
Mean ± std. dev.	--	0.9480 ± 0.02086	0.9483 ± 0.01044	0.9479 ± 0.006935
mse	--	0.02095	0.01057	0.007243
Iteration 2				
Mean ± std. dev.	--	0.9531 ± 0.02286	0.9506 ± 0.1011	0.9489 ± 0.006760
mse	--	0.02307	0.01011	0.006855
Iteration 3				
Mean ± std. dev.	--	0.9536 ± 0.02681	0.9506 ± 0.01087	0.9488 ± 0.006845
mse		0.02705	0.01089	0.006950
Theoretical std. dev.		0.03266	0.00833	--

have to use the procedure described in (71)-(73). In fact, we computed $\hat{S}_m^{-1/2}$ by the Levinson-Durbin algorithm, applied to the first column of $\hat{S}_m$. We used (67) and (69) to estimate $\hat{r}_v(\tau)$.

OIV-2: The factorization of $\hat{S}_v(z)$ was performed by computing the roots of $z^{nc}\hat{S}_v(z)$. All the roots outside the unit circle were reflected inside the unit circle, and the complete set of roots was then used to compute $\hat{C}^2(z)$. In this case we noticed that the filtering operation [by $G(q^{-1}) = 1/\hat{C}^2(q^{-1})$] introduced a transient that needed to be eliminated. To limit the duration of the transient we "contracted" the roots of the polynomial $\hat{C}^2(z)$ by replacing $\hat{C}^2(z) = 1 + \hat{c}_1'z + \cdots + \hat{c}_{nc}'z^{2nc}$ by $\hat{C}^2(z/\eta) = 1 + \hat{c}_1'\eta z + \cdots + \hat{c}_{2nc}'\eta^{2nc}z^{2nc}$, where $\eta = 0.99$. By construction, the roots of $\hat{C}^2(z)$ have maximum modulus of 1. To eliminate the effects of transients in $\hat{G}(q^{-1})y(t)$, the first 200 samples of the filtered data were discarded.

OIV-3: Implementation was very similar to OIV-2.

X. CONCLUSIONS

We presented a detailed analysis of the accuracy aspects of a general IV method for estimating the AR parameters of an ARMA process. The basic accuracy result (Theorem 1) is useful for evaluating the performance bounds for the various MYW-related estimation techniques discussed in the literature. See, for example, the discussion in [23,30].

More importantly, Theorem 1 can be used to investigate the existence of optimal IV methods. We derived a lower bound on the estimation accuracy of IV estimators and presented methods for achieving this bound.

The first method involved an optimal weighting matrix $Q = S^{-1}$ and letting the number m of instrumental variables increase to infinity. In this case the choice of the filter $G(q^{-1})$ becomes unimportant in the limit and we may set $G(q^{-1}) = 1$.

The second method involved an optimal filtering operation $G(q^{-1}) = 1/C^2(q^{-1})$. In this case the asymptotic bound is achieved for $m = na$, and the choice of the weighting matrix Q is unimportant.

The third method involved a filtering operation by $G(q^{-1}) = A(q^{-1})/C^2(q^{-1})$, which led to a very simple optimal weighting matrix $Q = I$. The best asymptotic accuracy is achieved for m tending to infinity.

Furthermore, we have shown that the optimal IV methods have the same (asymptotic) accuracy as the prediction error method (see Theorem 3).

The optimal IV methods suggest new algorithms for estimating the AR parameters of ARMA models. Note that these models require knowledge of certain quantities [such as $C(q^{-1})$ and $A(q^{-1})$] that are not available a priori. We have shown that replacing these quantities by their consistent estimates does not degrade the asymptotic estimation accuracy, and we described in detail three algorithms based on such replacement. These algorithms were shown to provide asymptotically efficient estimates of the AR parameters at a modest computational cost, compared to methods such as the maximum likelihood estimator. The performance of the proposed algorithms was illustrated by selected numerical examples.

Finally, we note that the optimal weighting matrix $Q = S^{-1}$ (required by the first method) can be estimated without explicit estimation of the MA parameters. This is convenient in some applications in which one needs only estimates of the AR parameters

APPENDIX A: PROOF OF THEOREM 3

Let us introduce the following notation:

$$r_k = E\left\{\frac{C(q^{-1})}{A(q^{-1})} e(t) \frac{1}{C(q^{-1})} e(t-k)\right\},$$

$$\tilde{R}_k = \begin{bmatrix} \tilde{r}_k \\ \tilde{r}_{k-1} \\ \vdots \\ \tilde{r}_{k-na+1} \end{bmatrix}. \tag{A.1}$$

It is straightforward to show that

$$\tilde{r}_k + a_1\tilde{r}_{k-1} + \cdots + a_{na}\tilde{r}_{k-na}$$
$$= E\left\{C(q^{-1})e(t)\frac{1}{C(q^{-1})} e(t-k)\right\} = 0$$
$$\text{for} \quad k \geq nc+1, \tag{A.2}$$

and hence

$$\tilde{R}_k = \underline{A}\tilde{R}_{k-1} \quad \text{for} \quad k \geq nc+1, \tag{A.3}$$

where $\underline{A}$ is the following companion matrix associated with the polynomial $A(z)$:

$$\underline{A} = \begin{bmatrix} -a_1 & -a_2 & \cdots & -a_{na} \\ 1 & \ddots & 0 & \vdots \\ 0 & & 1 & \end{bmatrix}. \tag{A.4}$$

It follows from (A.1)-(A.3) and (23) that

$$\tilde{P}_\infty^{-1} - \underline{A}\tilde{P}_\infty^{-1}\underline{A}^T = \frac{1}{\lambda^2}\left\{\sum_{i=nc}^{\infty} \tilde{R}_i\tilde{R}_i^T - \underline{A}\left[\sum_{i=nc}^{\infty} \tilde{R}_i\tilde{R}_i^T\right]\underline{A}^T\right\}$$

$$= \frac{1}{\lambda^2}\tilde{R}_{nc}\tilde{R}_{nc}^T.$$

In other words, $\tilde{P}_\infty^{-1}$ satisfies the following Lyapunov equation [see also (B.17)]:

$$\tilde{P}_\infty^{-1} - \underline{A}\tilde{P}_\infty^{-1}\underline{A}^T = \frac{1}{\lambda^2}\tilde{R}_{nc}\tilde{R}_{nc}^T. \tag{A.5}$$

Since $\underline{A}$ is a stability matrix, (A.5) has a *unique* solution (see, e.g., (20)). To show that $\tilde{P}_\infty = P_{PEM}$ it is thus sufficient and necessary to show that P_{PEM}^{-1} satisfies the same Lyapunov equation (A.5). We do this in the following steps. First note that

$$\underline{A}\psi_1(t) = \psi_1(t+1) - e(t)u_1, \tag{A.6}$$

where

$$u_1 = \underbrace{[1 \quad 0 \cdots 0]}_{na}^T, \tag{A.7}$$

and therefore

$$\underline{A}D_{11}A^T = D_{11} - \lambda^2 u_1 u_1^T, \tag{A.8}$$

where D_{11} is as defined in (31).

Next, we introduce

$$\underline{C} = \begin{bmatrix} -c_1 & \cdots & \cdots & -c_{nc} \\ 1 & & 0 & 0 \\ & \ddots & & \vdots \\ 0 & & 1 & \end{bmatrix} \tag{A.9}$$

and note that since $c_{nc} \neq 0$ the companion matrix $\underline{C}$ is nonsingular, and that

$$\underline{C}\psi_2(t) = \psi_2(t+1) - e(t)u_2, \tag{A.10}$$

where

$$u_2 = \underbrace{[1 \quad 0 \cdots 0]}_{nc}^T. \tag{A.11}$$

We can now write

$$\underline{A}D_{12}D_{22}^{-1}D_{12}^T A^T - \underline{A}D_{12}\underline{C}^T\left(\underline{C}D_{22}\underline{C}^T\right)^{-1}\underline{C}^T D_{12}^T \underline{A}^T$$

$$= \left(D_{12} - \lambda^2 u_1 u_2^T\right)\left(D_{22} - \lambda^2 u_2 u_2^T\right)^{-1}\left(D_{12}^T - \lambda^2 u_2 u_1^T\right), \tag{A.12}$$

where D_{12}, D_{21}, D_{22} are as defined in (31b). It follows from the matrix inversion lemma that

$$\left(D_{22} - \lambda^2 u_2 u_2^T\right)^{-1} = D_{22}^{-1} + \frac{\lambda^2 D_{22}^{-1} u_2 u_2^T D_{22}^{-1}}{1 - \lambda^2 u_2^T D_{22}^{-1} u_2}. \tag{A.13}$$

By using (A.8), A.12), and (A.13) we obtain after some straightforward but somewhat tedious calculations,

$$P_{PEM}^{-1} - \underline{A}P_{PEM}^{-1}\underline{A}^T = \lambda^2 \frac{\left(u_1 - D_{12}D_{22}^{-1}u_2\right)\left(u_1 - D_{12}D_{22}^{-1}u_2\right)^T}{1 - \lambda^2 u_2^T D_{22}^{-1} u_2}. \tag{A.14}$$

According to a well-known formula for the inverse of the covariance matrix of an AR process [27], we have

$$D_{22}^{-1}u_2 = \frac{1}{\lambda^2}\left\{\begin{bmatrix} 1 \\ c_1 \\ \vdots \\ c_{nc-1} \end{bmatrix} - c_{nc}\begin{bmatrix} c_{nc} \\ c_{nc-1} \\ \vdots \\ c_1 \end{bmatrix}\right\}. \tag{A.15}$$

To proceed we note the following properties of the covariance elements of D_{12}. Let

$$\gamma_k = E\left\{\frac{1}{A(q^{-1})} e(t) \frac{1}{C(q^{-1})} e(t - k)\right\}. \tag{A.16}$$

We have

$$\gamma_k + c_1\gamma_{k-1} + \cdots + c_{nc}\gamma_{k-nc} = E\left\{\frac{C(q^{-1})}{A(q^{-1})}\, e(t)\frac{1}{C(q^{-1})}\, e(t-k)\right\}$$

$$= \tilde{r}_k \quad \text{for all} \quad k \qquad \text{(A.17)}$$

and

$$\gamma_k + c_1\gamma_{k+1} + \cdots + c_{nc}\gamma_{k+nc} = E\left\{\frac{1}{A(q^{-1})}\, e(t)e(t-k)\right\}$$

$$= \begin{cases} \lambda^2, & k = 0 \\ 0, & k < 0. \end{cases} \qquad \text{(A.18)}$$

If, similarly to (A.1), we introduce

$$\Gamma_k = \begin{bmatrix} \gamma_k \\ \gamma_{k-1} \\ \vdots \\ \gamma_{k-na+1} \end{bmatrix},$$

then from (A.15)-(A.18) we have

$$D_{12}D_{22}^{-1}u_2 = [\Gamma_0 \quad \Gamma_1 \quad \cdots \quad \Gamma_{nc-1}]\left\{\begin{bmatrix} 1 \\ c_1 \\ \vdots \\ c_{nc-1} \end{bmatrix} - c_{nc}\begin{bmatrix} c_{nc} \\ c_{nc-1} \\ \vdots \\ c_1 \end{bmatrix}\right\}\Bigg/\lambda^2$$

$$= \left[\lambda^2 u_1 - c_{nc}\Gamma_{nc} - c_{nc}\left(\tilde{R}_{nc} - \Gamma_{nc}\right)\right]/\lambda^2$$

$$= u_1 - \frac{c_{nc}}{\lambda^2}\,\tilde{R}_{nc},$$

which gives

$$u_1 - D_{12}D_{22}^{-1}u_2 = \frac{c_{nc}}{\lambda^2}\,\tilde{R}_{nc}.$$

To evaluate the denominator of the right-hand side of (A.14), we use (A.15) to obtain

$$1 - \lambda^2 u_2^T D_{22}^{-1} u_2 = 1 - \left(1 - c_{nc}^2\right) = c_{nc}^2.$$

It follows that the right-hand side of (A.14) reduces to $(1/\lambda^2)\tilde{R}_{nc}\tilde{R}_{nc}^T$, which is precisely the right-hand side of (A.5). We have shown that $\tilde{P}^{-1}$ and P_{PEM}^{-1} obey the same Lyapunov equation, and therefore $\tilde{P}_{\infty}^{-1} = P_{PEM}^{-1}$.

APPENDIX B: CONVERGENCE OF $\tilde{P}_m$

In this appendix we consider the convergence as $m \to \infty$ of the inverse of the optimal error covariance matrix

$$\tilde{P}_m^{-1} = R_m^T S_m^{-1} R_m, \tag{B.1}$$

where R_m and S_m are defined by (11) and (16), respectively. We start by introducing the following notation:

$$\bar{r}_k = E\{y(t)G(q^{-1})y(t - k)\},$$

$$\bar{R}_k = \begin{bmatrix} \bar{r}_k \\ \bar{r}_{k-1} \\ \vdots \\ \bar{r}_{k-na+1} \end{bmatrix}. \tag{B.2}$$

Note that

$$\begin{aligned} &\bar{r}_k + a_1\bar{r}_{k-1} + \cdots + a_{na}\bar{r}_{k-na} \\ &\quad = E\{A(q^{-1})y(t)G(q^{-1})y(t - k)\} = 0, \quad k \geq nc + 1. \end{aligned} \tag{B.3}$$

If we let $\underline{A}$ be the companion matrix defined in (A.4), then (B.3) implies that

$$\bar{R}_k = \underline{A}\bar{R}_{k-1}, \quad k \geq nc + 1. \tag{B.4}$$

Let us also introduce

$$\begin{aligned} \bar{y}(t) &= C(q^{-1})G(q^{-1})y(t), \\ \psi_m^T &= E\{\bar{y}(t)\,[\bar{y}(t - 1) \cdots \bar{y}(t - m)]\}, \\ \alpha_m^2 &= E\{\bar{y}^2(t)\} - \psi_m^T S_m^{-1}\psi_m. \end{aligned} \tag{B.5}$$

We can now state the following result.

Lemma B.1

Consider the sequence of matrices $\tilde{P}_m^{-1}$, $m = 1, 2, \ldots$, defined by (B.1). The following Lyapunov-type equation holds true:

$$\tilde{P}_{m+1}^{-1} - A\tilde{P}_m^{-1}\underline{A}^T = \frac{1}{\alpha_m^2}\left(\bar{R}_{nc} - \underline{A}R_m^T S_m^{-1}\psi_m\right)\left(\bar{R}_{nc} - \underline{A}R_m^T S_m^{-1}\psi_m\right)^T,$$

$$m = 1, 2, \ldots . \tag{B.6}$$

Proof. First note that according to (B.4),

$$\begin{aligned} R_{m+1}^T &= E\left\{\phi(t) z_{m+1}^T(t)\right\} \\ &= \left[\bar{R}_{nc}, \bar{R}_{nc+1}, \ldots, \bar{R}_{nc+m}\right] = \left[\bar{R}_{nc}, AR_m^T\right]. \end{aligned}$$

Next, we have

$$\begin{aligned} S_{m+1}^{-1} &= \begin{bmatrix} E\{\bar{y}^2(t)\} & \psi_m^T \\ \psi_m & S_m \end{bmatrix}^{-1} \\ &= \begin{bmatrix} 0 & 0 \\ 0 & S_m^{-1} \end{bmatrix} + \frac{1}{\alpha_m^2}\begin{bmatrix} -1 \\ S_m^{-1}\psi_m \end{bmatrix}\left[-1, \psi_m^T S_m^{-1}\right]. \end{aligned}$$

Therefore, we can write

$$\begin{aligned} \tilde{P}_{m+1}^{-1} &= R_{m+1}^T S_{m+1}^{-1} R_{m+1} \\ &= \underline{A}R_m^T S_m^{-1} R_m \underline{A}^T + \frac{1}{\alpha_m^2}\left(\bar{R}_{nc} - \underline{A}R_m^T S_m^{-1}\psi_m\right)\left(\bar{R}_{nc} - \underline{A}R_m^T S_m^{-1}\psi_m\right)^T, \end{aligned}$$

which concludes the proof.

Next, we study the limit as $m \to \infty$ of the right-hand side of (B.6).

Lemma B.2

Let $m \to \infty$. Then, under assumptions A1-A3,

$$\alpha_m^2 \to \lambda^2, \tag{B.7a}$$

$$\bar{R}_{nc} - \underline{A}R_m^T S_m^{-1}\psi_m \to E\left\{\phi(t)\frac{1}{C(q^{-1})}e(t - nc - 1)\right\} \triangleq \tilde{R}_{nc}. \tag{B.7b}$$

Proof. Define

$$\sum_{i=0}^{\infty} h_i q^{-i} \triangleq \frac{A(q^{-1})}{C^2(q^{-1})G(q^{-1})} \qquad h_0 = 1. \tag{B.8}$$

Due to assumptions A1 and A3,

$$|h_k| < c\mu^k, \tag{B.9}$$

where c is a constant and $0 < \mu < 1$ is the maximum modulus of the zeros of $C^2(a^{-1})G(q^{-1})$. Now $A(q^{-1})\overline{y}(t) = C^2(q^{-1})G(q^{-1})e(t)$, so that for large enough m we can write

$$\overline{y}(t) + h_1\overline{y}(t - 1) + \cdots + h_m\overline{y}(t - m) + 0(\mu^{m+1}) = e(t). \tag{B.10}$$

It follows from (B.10) that

$$\psi_m = S_m \begin{bmatrix} h_1 \\ \vdots \\ h_m \end{bmatrix} + 0(\mu^{m+1}). \tag{B.11}$$

Hence

$$\underline{A}R_m^T S_m^{-1} \psi_m = -\underline{A}R_m^T \begin{bmatrix} h_1 \\ \vdots \\ h_m \end{bmatrix} + 0(m\mu^{m+1}), \tag{B.12}$$

and

$$\psi_m^T S_m^{-1} \psi_m = -\psi_m^T \begin{bmatrix} h_1 \\ \vdots \\ h_m \end{bmatrix} + 0(m\mu^{m+1}). \tag{B.13}$$

Consider first (B.12). We have

$$R_m^T \begin{bmatrix} h_1 \\ \vdots \\ h_m \end{bmatrix} = E\left\{\phi(t)G(q^{-1})\left[\sum_{i=1}^{m} h_i y(t - nc - i)\right]\right\}$$

$$= E\left\{\phi(t)G(q^{-1})\left[\sum_{i=0}^{\infty} h_i y(t - nc - i) - y(t - nc) + 0(\mu^{m+1})\right]\right\}$$

$$= E\left\{\phi(t)\,\frac{1}{C(q^{-1})}\,e(t - nc)\right\} - E\{\phi(t)G(q^{-1})y(t - nc)\}$$

$$+ 0(\mu^{m+1}). \tag{B.14}$$

Further straightforward calculations give

$$\underline{A}\phi(t) = \phi(t + 1) - C(q^{-1})e(t)u_1, \qquad u_1 = \begin{bmatrix} 1 \\ 0 \\ \vdots \\ 0 \end{bmatrix} \text{na}. \tag{B.15}$$

Combining (B.14) and (B.15), we obtain

$$\underline{A}R_m^T\begin{bmatrix} h_1 \\ \vdots \\ h_m \end{bmatrix} = \tilde{R}_{nc} - c_{nc}\lambda^2 u_1 - \bar{R}_{nc} + c_{nc}\lambda^2 u_1 + 0(\mu^{m+1})$$

$$= \tilde{R}_{nc} - \bar{R}_{nc} + 0(\mu^{m+1}). \tag{B.16}$$

This equation, together with (B.12), implies (B.7b). Next consider (B.13). We have

$$\psi_m^T\begin{bmatrix} h_1 \\ \vdots \\ h_m \end{bmatrix} = E\left\{\bar{y}(t)\left[\sum_{i=1}^{m} h_i \bar{y}(t - i)\right]\right\}$$

$$= E\left\{\bar{y}(t)\left[\sum_{i=0}^{\infty} h_i \bar{y}(t - i) - \bar{y}(t) + 0(\mu^{m+1})\right]\right\}$$

$$= E\{\bar{y}(t)e(t)\} - E\{\bar{y}^2(t)\} + 0(\mu^{m+1})$$

$$= \lambda^2 - E\{\bar{y}^2(t)\} + 0(\mu^{m+1}),$$

which, together with (B.5) and (B.13), proves (B.7a). This concludes the proof of Lemma B.2.

It is now straightforward to evaluate $\tilde{P}_\infty \triangleq \lim_{m\to\infty} \tilde{P}_m$. The limit exists since $\tilde{P}_m \geq \tilde{P}_{m+1} > 0$ (see Lemma 2). Furthermore, it follows from Lemmas B.1 and B.2 that $\tilde{P}_\infty^{-1}$ satisfies

$$\tilde{P}_\infty^{-1} - \underline{A}\tilde{P}_\infty^{-1}\underline{A}^T = \frac{1}{\lambda^2} \tilde{R}_{nc}\tilde{R}_{nc}^T. \tag{B.17}$$

As is well known, under the given assumptions the solution of (B.17) is unique and is given by

$$\tilde{P}_\infty^{-1} = \frac{1}{\lambda^2} \sum_{k=0}^{\infty} \underline{A}^k\tilde{R}_{nc}\tilde{R}_{nc}^T(\underline{A}^T)^k.$$

In Appendix A we have shown that $R_k = \underline{A}R_{k-1}$. Therefore,

$$\tilde{P}_\infty = \lambda^2 \left\{ \begin{bmatrix} \tilde{R}_{nc} & \tilde{R}_{nc+1} & \cdots \end{bmatrix} \begin{bmatrix} \tilde{R}_{nc}^T \\ \tilde{R}_{nc+1}^T \\ \vdots \end{bmatrix} \right\}^{-1}$$

which is precisely equation (23).

APPENDIX C: PROOF OF LEMMA 1

Note that we can write

$$S_{m+1} = \begin{bmatrix} S_m & \psi_m \\ \psi_m^T & \sigma \end{bmatrix},$$

$$\psi_m = E\left\{C(q^{-1})z_m(t)C(q^{-1})G(q^{-1})y(t - nc - m - 1)\right\},$$

$$\sigma = E\{C(q^{-1})G(q^{-1})y(t)\}^2,$$

and

$$R_{m+1} = \begin{bmatrix} R_m \\ \phi_m^T \end{bmatrix},$$

$$\phi_m = E\{\phi(t)G(q^{-1})y(t - nc - m - 1)\}.$$

Therefore,

$$\tilde{P}_{m+1}^{-1} = R_{m+1}^T S_{m+1}^{-1} R_{m+1}$$

$$= \begin{bmatrix} R_m^T & \phi_m \end{bmatrix} \left\{ \begin{bmatrix} I \\ 0 \end{bmatrix} S_m^{-1} [I \quad 0] + \alpha_m \begin{bmatrix} S_m^{-1}\psi_m \\ -1 \end{bmatrix} \begin{bmatrix} \psi_m^T S_m^{-1}, & -1 \end{bmatrix} \right\} \begin{bmatrix} R_m \\ \phi_m^T \end{bmatrix}$$

$$= \tilde{P}_m^{-1} + \alpha_m \left[\phi_m - R_m^T S_m^{-1} \psi_m\right]\left[\phi_m - R_m^T S_m^{-1} \psi_m\right]^T, \qquad \text{(C.1)}$$

where

$$\alpha_m = \frac{1}{\alpha - \psi_m^T S_m^{-1} \psi_m}.$$

Since $S_m > 0$ for all m, we have $0 < \alpha_m < \infty$. This proves the order order relation $\tilde{P}_m \geq \tilde{P}_{m+1}$. Furthermore, it follows from (C.1) that

$$\left\{\tilde{P}_m = \tilde{P}_{m+1},\ m \geq na\right\} \leftrightarrow \left\{\phi_m = R_m^T S_m^{-1} \psi_m,\ m \geq na\right\}. \qquad \text{(C.2)}$$

To obtain (37) from (C.2) notice first that

$$x_m = E\left\{\frac{C^2(q^{-1})G(q^{-1})}{A(q^{-1})} \begin{bmatrix} e(t-1) \\ \vdots \\ e(t-m) \end{bmatrix} A(q)\ \frac{C^2(q^{-1})G(q^{-1})}{A(q^{-1})}\ e(t-m-1)\right\}$$

$$= E\left\{C(q^{-1})G(q^{-1}) \begin{bmatrix} y(t-1) \\ \vdots \\ y(t-m) \end{bmatrix}\right.$$

$$\times \{C(q^{-1})G(q^{-1})[y(t-1) \cdots y(t-m)]$$

$$\left.\times\ a + C(q^{-1})G(q^{-1})y(t-m-1)\}\right\}$$

$$= S_m a + \psi_m, \qquad \text{(C.3)}$$

where

$$a = [0, \ldots, 0, a_{na}, \ldots, a_1]^T.$$

Next introduce

$$\bar{r}_k = E\{y(t)G(q^{-1})y(t-k)\}; \qquad \bar{R}_k = \begin{bmatrix} \bar{r}_k \\ \bar{r}_{k-1} \\ \vdots \\ \bar{r}_{k-na+1} \end{bmatrix} \tag{C.4}$$

and note that

$$\bar{r}_k + a_1\bar{r}_{k-1} + \cdots + a_{na}\bar{r}_{k-na} = E\{C(q^{-1})e(t)G(q^{-1})y(t-k)\} = 0,$$
$$k \geq nc + 1, \tag{C.5}$$

and, therefore, that

$$\bar{R}_k + a_1\bar{R}_{k-1} + \cdots + a_{na}\bar{R}_{k-na} = 0 \quad \text{for} \quad k \geq nc + na.$$

It follows from (C.3) that

$$\phi_m - R_m^T S_m^{-1} \psi_m = \phi_m + R_m^T a - R_m^T S_m^{-1} x_m.$$

However, from (C.1) and (C.2) we have

$$\phi_m + R_m^T a = \bar{R}_{nc+m} + \left[\bar{R}_{nc} \cdots \bar{R}_{nc+m-1}\right] \begin{bmatrix} 0 \\ \vdots \\ 0 \\ a_{na} \\ \vdots \\ a_1 \end{bmatrix}$$

$$= \bar{R}_{nc+m} + a_1\bar{R}_{nc+m-1} + \cdots + a_{na}\bar{R}_{nc+m-na}$$

$$= 0 \quad \text{for} \quad m \geq na.$$

Hence (C.2) reduces to (37) and the proof is completed.

APPENDIX D: THE BEST POSITIVE-DEFINITE APPROXIMATION OF A SYMMETRIC MATRIX

Let A be an $m \times m$ symmetric matrix. Let $\lambda_1 \geq \lambda_2 \geq \cdots \geq \lambda_m$ be its eigenvalues and $v_1, \ldots, v_m$ be the corresponding eigenvectors. We have the following result.

Lemma D.1

Let C be the class of positive definite $m \times m$ matrices with eigenvalues larger than or equal to a given (small) positive number ϵ. Then

$$\inf_{B \in C} \|A - B\| = \left[(\lambda_{n+1} - \epsilon)^2 + \cdots + (\lambda_m - \epsilon)^2\right]^{1/2}, \tag{D.1}$$

where $\|A\| = [\operatorname{tr} AA^T]^{1/2} = \Sigma_i \Sigma_j a_{ij}^2{}^{\;1/2}$ denotes the Euclidean norm and $\lambda_{n+1}, \ldots, \lambda_m$ are the eigenvalues of A that are smaller than ϵ, that is,

$$\begin{aligned} &\lambda_k \geq \epsilon, \quad k = 1, \ldots, n, \\ &\lambda_k < \epsilon \quad k = n + 1, \ldots, m. \end{aligned} \tag{D.2}$$

Furthermore, the infimum is attained for

$$B = V \begin{bmatrix} \lambda_1 & & & & & 0 \\ & \ddots & & & & \\ & & \lambda_n & & & \\ & & & \epsilon & & \\ & 0 & & & \ddots & \\ & & & & & \epsilon \end{bmatrix} V^T, \tag{D.3}$$

with $V = [v_1, \ldots, v_m]$.

Proof. We have

$$\|A - B\|^2 = \|V^T(A - B)V\|^2 = \sum_{i=1}^{m} (\lambda_i - c_{ii})^2 + \sum_{\substack{i,j=1 \\ i \neq j}}^{m} c_{ij}^2,$$

where c_{ij} is the i, j-element of

$$C = V^T BV. \tag{D.4}$$

Clearly, C has the same eigenvalues as B. Thus we can write

$$\|A - B\|^2 \geq \sum_{i=1}^{m} (\lambda_i - c_{ii})^2 \geq \sum_{i=n+1}^{m} (\lambda_i - c_{ii})^2 \geq \sum_{i=n+1}^{m} (\lambda_i - \epsilon)^2,$$

where the equalities hold if

$$\begin{aligned} &c_{ij} = 0, \quad i \neq j; \quad c_{ii} = \lambda_i, \quad i = 1, \ldots, n; \\ &c_{ii} = \epsilon, \quad i = n + 1, \ldots, m. \end{aligned} \tag{D.5}$$

By inserting (D.5) into (D.4), we readily obtain (D.3).

APPENDIX E: A RECURSIVE QR ALGORITHM FOR SOLVING (6)

Let us rewrite (6) as

$$L_m \hat{\theta}_m = \ell_m, \tag{E.1}$$

where

$$L_m = \hat{S}_m^{-1/2} \sum_{t=1}^{N} z_m(t)\phi(t)^T, \tag{E.2}$$

$$\ell_m = \hat{S}_m^{-1/2} \sum_{t=1}^{N} z_m(t)y(t). \tag{E.3}$$

Let L_m be factored as

$$L_m = O_m T_m, \tag{E.4}$$

where O_m is an orthogonal matrix and T_m an upper triangular matrix. Then $\hat{\theta}_m$ can be computed by back-substitution from

$$T_m \hat{\theta}_m = O_m^T \ell_m. \tag{E.5}$$

Consider now the situation for m + 1. Determine first α and β in

$$\hat{S}_{m+1}^{-1/2} = \begin{bmatrix} \hat{S}_m^{-1/2} & 0 \\ \beta^T & \alpha \end{bmatrix}, \tag{E.6}$$

and then

$$L_{m+1} = \hat{S}_{m+1}^{-1/2} R_{m+1} = \begin{bmatrix} L_m \\ \lambda^T \end{bmatrix}. \tag{E.7}$$

We have

$$L_{m+1} = \begin{bmatrix} O_m & 0 \\ 0 & 1 \end{bmatrix} \begin{bmatrix} T_m \\ \gamma^T \end{bmatrix}. \tag{E.8}$$

So the problem of factorizing L_{m+1} reduces to the factorization of $\begin{bmatrix} T_m \\ \gamma^T \end{bmatrix}$. In this last matrix only the last row, γ^T, needs to be made zero. The computations needed are clearly simpler than if the matrix had been full. Let $\bar{O}_{m+1}$ be an othogonal matrix such that $\bar{O}_{m+1}\begin{bmatrix} T_m \\ \gamma^T \end{bmatrix}$ is an upper-triangular matrix $\triangleq T_{m+1}$, or

$$L_{m+1} = \underbrace{\begin{bmatrix} O_m & 0 \\ 0 & 1 \end{bmatrix} \bar{O}^T_{m+1}}_{O_{m+1}} T_{m+1} = O_{m+1} T_{m+1}. \tag{E.9}$$

Finally, we have

$$\ell_{m+1} = \hat{S}^{-1/2}_{m+1} \sum_{t=1}^{N} z_{m+1}(t)\phi(t)^T = \begin{bmatrix} \ell_m \\ \rho \end{bmatrix}. \tag{E.10}$$

The estimate $\hat{\theta}_{m+1}$ is computed from

$$T_{m+1}\hat{\theta}_{m+1} = O^T_{m+1}\ell_{m+1} = O_{m+1}\begin{bmatrix} O^T_m & 0 \\ 0 & 1 \end{bmatrix}\begin{bmatrix} \ell_m \\ \rho \end{bmatrix}, \tag{E.11}$$

or

$$T_{m+1}\hat{\theta}_{m+1} = \bar{O}_{m+1}\begin{bmatrix} O^T_m \ell_m \\ \rho \end{bmatrix}. \tag{E.12}$$

ACKNOWLEDGMENTS

The authors gratefull acknowledge the contribution of Dr. J. O. Smith, who provided the simulation results for Section IX. The work of B. Friedlander was supported by the Army Research office under Contract DAAG29-83-C-0027.

REFERENCES

1. J. A. CADZOW, "ARMA Modeling of Time Series," *IEEE Trans. Pattern Anal. Mach. Intelligence PAMI-4*, No. 2, 124-128, March 1982.

2. J. A. CADZOW, "Spectral Estimation: An Overdetermined Rational Model Equation Approach," *Proc. IEEE 70*, No. 4, 907-939, September 1982.

3. Y. T. CHAN and R. P. LANGFORD, "Spectral Estimation *via* the High-Order Yule-Walker Equation," *IEEE Trans. Acoust. Speech Signal Process. ASSP-30*, No. 5, 689-698, October 1982.

4. B. Friedlander, "The Overdetermined Recursive Instrumental Variable Method," *IEEE Trans. Autom. Control AC-29*, No. 4, 353-356, April 1984.

5. J. A. CADZOW, "High Performance Spectral Estimation—A New ARMA Method," *IEEE Trans. Acoust. Speech Signal Process. ASSP-28*, No. 5, 524-529, October 1980.

6. S. M. KAY, "A New ARMA Spectral Estimator," *IEEE Trans. Acoust. Speech Signal Process. ASSP-28*, No. 5, 585-588, October 1980.

7. W. GERSCH, "Estimation of the Autoregressive Parameters of a Mixed Autoregressive Moving-Average Time Series," *IEEE Trans. Autom. Control AC-15*, 583-588 (1970).

8. B. FRIEDLANDER, "Instrumental Variable Methods for ARMA Spectral Estimation," *IEEE Trans. Acoust. Speech Signal Process. ASSP-31*, No. 2, 404-415, April 1983.

9. B. FRIEDLANDER and B. PORAT, "The Modified Yule-Walker Method of ARMA Specral Estimation," *IEEE Trans. Aerosp. Electron. Syst. AES-20*, No. 2, 158-173, March 1984.

10. P. STOICA, "Generalized Yule-Walker Equations and Testing the Orders of Multivariate Time Series," *Int. J. Control 37*, No. 5, 1159-1166 (1983).

11. T. SÖDERSTRÖM and P. STOICA, "Comparison of Some Instrumental Variable Methods—Consistency and Accuracy Aspects," *Automatica 17*, No. 1, 101-115 (1981).

12. T. SÖDERSTRÖM and P. STOICA, "Instrumental Variable Methods for System Identification," Springer-Verlag, Berlin and New York, 1983.

13. P. STOICA and T. SÖDERSTRÖM, "Optimal Instrumental Variable Estimation and Approximate Implementation," *IEEE Trans. Autom. Control AC-28*, No. 7, 757-772, July 1983.

14. P. STOICA, "On a Procedure for Testing the Orders of Time Series," *IEEE Trans. Autom. Control AC-26*, 572-573 (1981).

15. T. SÖDERSTRÖM, "Ergodicity Results for Sample Covariances," *Probl. Control Inf. Theory 4*, 1331-1338 (1975).

16. K. J. ASTRÖM and T. SÖDERSTRÖM, "Uniqueness of the Maximum Likelihood Estimates of the Parameters of an ARMA Model," *IEEE Trans. Autom. Control AC-19*, 769-773 (1974).

17. G. E. P. BOX and G. M. JENKINS, "Time Series Analysis—Forecasting and Control," Holden-Day, San Francisco, California, 1976.

18. L. LJUNG, "On Consistency of Prediction Error Identification Methods," *in* "System Identification, Advances and Case Studies," (R. K. Mehra and D. G. Lainiotis, eds.), Academic Press, New York, 1976.

19. P. E. CAINES and L. LJUNG, "Prediction Error Estimates: Asymptotic Normality and Accuracy," *Proc. IEEE Conf. Decision Control* (1976).

20. T. KAILATH, "Linear Systems," Prentice-Hall, Englewood Cliffs, New Jersey, 1980.

21. T. SÖDERSTRÖM, Unpublished Notes on IV Estimation of AR Parameters of ARMA Processes, 1983.

22. C. R. RAO, "Linear Statistical Inference and Its Applications," Wiley, New York, 1973.

23. B. FRIEDLANDER and K. C. SHARMAN, "Performance Analysis of the Modified Yule-Walker Estimator," *IEEE Trans. Acoust. Speech Signal Process. ASSP-33*, No. 3, 719-725, June 1985.

24. P. STOICA, B. FRIEDLANDER, and T. SÖDERSTRÖM, "Optimal Instrumental Variable Multistep Algorithms for Estimation of the AR Parameters of an ARMA Process," Technical Report 5498-04, Systems Control Technology, Inc., Palo Alto, California, May 1984; also, *Proc. IEEE Conf. Decision Control, Florida*, December, 1985.

25. C. L. LAWSON and R. J. HANSON, "Solving Least-Squares Problems," Prentice-Hall, Englewood Cliffs, New Jersey, 1974.

26. L. LJUNG, "Convergence Analysis of Parametric Identification Methods," *IEEE Trans. Autom. Control AC-23*, 770-783, October 1978.

27. T. KAILATH, A. VIEIRA, and M. MORF, "Inverses of Toeplitz Operators, Innovations and Orthogonal Polynomials," *SIAM Rev. 20*, 106-110, January 1978.

28. K. J. ASTRÖM, "Introduction to Stochastic Control Theory," Academic Press, New York, 1970.

29. M. PAGANO, "Estimation of Models of Autoregressive Signal Plus Noise," *Ann. Stat. 2*, No. 1, 99-108 (1974).

30. P. STOICA, T. SÖDERSTRÖM, and B. FRIEDLANDER, "Optimal Instrumental Variable Estimates of the AR Parameters of an ARMA Process," *IEEE Trans. Autom. Control AC-30*, No. 11, 1066-1075, November 1985.

31. L. P. HANSEN, "Large Sample Properties of Generalized Method of Moments Estimators," *Econometrica 50*, No. 4, 1029-1054, July 1982.

32. F. HAYASHI and C. SIMS, "Nearly Efficient Estimation of Time Series Models with Predetermined, but not Exogenous, Instruments," *Econometrica 51*, No. 3, 783-798, May 1983.

33. L. P. HANSEN and Th. SARGENT, "Instrumental Variable Procedures for Estimating Linear Rational Expectations Models," *J. Monetary Econ. 9*, 263-296 (1982).

34. G. T. WILSON, "Factorization of the Covariance Generating Function of a Pure Moving-Average Process," *SIAM J. Numer. Anal. 6*, 1-7 (1969).

35. V. KUČERA, "Discrete Linear Control—The Polynomial Equation Approach," Wiley, New York, 1979.

36. B. FRIEDLANDER, "A Lattice Algorithm for Factoring the Spectrum of a Moving-Average Process," *IEEE Trans. Autom. Control*, No. 11, 1051-1055, November 1983.

37. L. LJUNG and P. E. CAINES, "Asymptotic Normality of Prediction Error Estimators for Approximate System Models," *Stochastics 3*, 29-46 (1979).

Continuous and Discrete Adaptive Control

G. C. GOODWIN
R. MIDDLETON

*Department of Electrical and Computer Engineering
The University of Newcastle
New South Wales, 2308 Australia*

I. INTRODUCTION

An adaptive controller is a control law that is, in principle, capable of initially tuning itself and of retuning itself should the process characteristics subsequently change. The potential benefits arising from the use of this type of algorithm include higher performance, better energy efficiency, higher productivity, improved quality control, and enhanced safety margin.

There has recently been increased interest in adaptive control. This is a result of two main factors: (a) the availability of cheap and reliable computer hardware on which the algorithms can be implemented and (b) a greater understanding of the theory of adaptive control.

Our objective in this article is to describe a particular class of adaptive control laws. We will endeavor to place emphasis on robust algorithms. By robust we mean insensitivity to finite accuracy sampling and finite word length computation, insensitivity to whether the disturbances are modeled as stochastic processes or deterministic signals, equivalence between continuous and discrete designs with fast sampling, insensitivity to model accuracy, and downward compatibility to frequently used structures including PID controllers.

For simplicity we will restrict attention to the single-input single-output case. Natural extensions of some of the algorithms to the multi-input, multi-output case are available elsewhere in the literature [1,2].

We shall adopt the certainty equivalent approach [1,3,4] to the design of adaptive controllers. This approach involves two stages:

(i) estimation of the system parameters without regard for what they will be used; and

(ii) use of the estimated parameters in the control system design as if they were the true system parameters.

We shall first describe models for linear systems. We will then describe a class of algorithms for parameter estimation and a class of algorithms for control law synthesis. Finally, we will show how adaptive control laws can be designed by combining a parameter estimator with a control law synthesis procedure.

A key feature of our development is that we treat continuous and discrete systems within a unified framework. In particular, we present a unified convergence theory and describe a class of discrete-time adaptive control laws applicable to continuous-time systems with rapid sampling. The corresponding continuous-time adaptive control laws are then simply obtained as a limiting case when the sampling period tends to zero.

II. A CLASS OF PLANT MODELS

We will initially consider the noise-free case. We will later show how the model can be extended to include the effects of noise and disturbances. We assume that the underlying process is continuous in nature and can be described by a set of linear ordinary differential equations:

$$\frac{d}{dt} x(t) = A'x(t) + B'u(t), \tag{1}$$

$$y(t) = C'x(t). \tag{2}$$

In those cases wehre the control law is implemented digitally, we will assume that the plant output is measured by a sample hold circuit and that the plant input is implemented via a digital-to-analog converter with zero-order hold. The discrete-time model corresponding to (1) and (2) then becomes

$$qx(t) = F'x(t) + G'u(t), \tag{3}$$

$$y(t) = H'x(t), \tag{4}$$

where q is the forward shift operator,

$$F' = e^{A'\Delta}, \tag{5}$$

$$G' = (A')^{-1}(e^{A'\Delta} - I)B' \quad \text{(assuming } A' \text{ nonsingular)}, \tag{6}$$

$$H' = C', \tag{7}$$

and Δ is the sample period.

The model (3), (4) will be called a *shift operator model*. Models of this type have been used almost exclusively to describe sampled data systems. However, a disadvantage of the model (3), (4) is that with rapid sampling the eigenvalues of F' tend to cluster around the point $1 + j0$. This can lead to numerical difficulties since all of the dynamic information is encoded in the difference between the eigenvalues and 1. This suggests that it should be beneficial to shift the origin to the point 1 + j0. This is achieved by the *delta operator* defined as follows:

$$\delta = (q - 1)/\Delta. \qquad (8)$$

The notion of the delta operator has been known for some time. For example, it is defined in [5] and it is used in [6] as a way of motivating Z transforms and in [7] to relate continuous and discrete designs. More recently, it has been recognized that the delta operator gives enhanced numerical properties in the implementation of digital algorithms [8,9,10,11]. Since the delta operator simply represents a shift in origin, it is a straightforward matter to translate the existing theory of discrete estimation and control to cover this case. We shall see later that many advantages result from the use of the delta operator, including the fact that the equivalent continuous-time algorithms and theory are obtained simply by letting the sampling period Δ tend to zero.

Using the delta operator, (3) becomes

$$\delta x(t) = \overline{F}x(t) + \overline{G}u(t), \qquad (9)$$

where

$$\overline{F} = (F' - I)/\Delta, \quad \overline{G} = G'/\Delta.$$

We stress that models based on the delta operator provide an exact representation of the sampled response of the system. Of course, δ looks like an Euler approximation to a derivative. This implies that δ (in discrete time) and d/dt (in continouous time) have a similar (though not identical) physical interpretation. This, in turn, means that operations in the continuous and discrete domains have roughly equivalent significance. This is frequently helpful in gaining insight in the discrete case. For example, it is often forgotten that shift operator models of the type given in (3), (4) depend implicitly on high-order differences between successive outputs (up to order n). This is made explicit in the delta formulation, making it easy to detect those operations that would involve "near differentiation" of the data.

For the purposes of adaptive control it is necessary to give an explicit parametrization of the matrices $\overline{F}$, $\overline{G}$, and H'. There are two alternatives: (a) use a physically based model or (b) use a canonical structure. The former approach has the advantage that certain parameters may be known and fixed. However, the system response is, in general, a nonlinear function of the parameters leading to more complex estimation algorithms. If the latter approach is adopted, then the structure can be chosen so as to simplify the algorithm. For example, we will see below that the choice of observer state-space model [1] simplifies the parameter estimator since the model is linear (in a certain sense) in the parameters. For the single-input, single-output case, the observer form is as in (4), (9) with

$$\overline{F} = \begin{bmatrix} -a_{n-1} & 1 & & \\ \vdots & & \ddots & \\ & & & 1 \\ -a_0 & & 0 & 0 \end{bmatrix}, \qquad \overline{G} = \begin{bmatrix} b_{n-1} \\ \vdots \\ b_0 \end{bmatrix}, \tag{10}$$

$$H' = [1 \quad 0 \quad \cdots \quad 0]. \tag{11}$$

Substituting (10), (11) into (9), (4) and successively eliminating the states by back substitution leads to the following model:

$$A(\delta)y(t) = B(\delta)u(t), \tag{12}$$

where

$$A(\delta) = \delta^n + a_{n-1}\delta^{n-1} + \cdots + a_0, \tag{13}$$

$$B(\delta) = b_{n-1}\delta^{n-1} + \cdots + b_0. \tag{14}$$

The model (12) is usually called a *left matrix fraction description* [12] and is simply a compact way of writing the model (10), (11).

To complete the model (12) we need to add a description of noise and disturbances. We do this by augmenting the model as follows:

$$A(\delta)y(t) = B(\delta)u(t) + \eta(t) + d(t), \tag{15}$$

where $\eta(t)$ denotes noise and random disturbances and $d(t)$ denotes deterministic disturbances. To retain as much generality as possible, the noise term η is simply modeled by a filter, that is,

$$G(\delta)\eta(t) = F(\delta)\epsilon(t), \tag{16}$$

where $\epsilon(t)$ is small in some sense. We shall further require that F and G be monic and stable. A special case of (16) is when η is a stationary stochastic process, in which case F/G denotes the stable/minimum phase spectral factor of the spectral density of η and ϵ denotes the innovations sequence.

The disturbance d is assumed to be purely deterministic and to have a finite-dimensional model of the form of (9). This model will, in general, be observable but *not* controllable. Transforming to the model format (12), d(t) is described by

$$D(\delta)d(t) = 0, \tag{17}$$

with $D(\delta)$ monic.

For example, if d(t) is a sine wave,

$$d(t) = A\sin(\omega_0 t + \phi), \tag{18}$$

then an approapriate continuous-time model is

$$\frac{d}{dt}\begin{bmatrix} x_1(t) \\ x_2(t) \end{bmatrix} = \begin{bmatrix} 0 & 1 \\ -\omega_0^2 & 0 \end{bmatrix}\begin{bmatrix} x_1(t) \\ x_2(t) \end{bmatrix}, \tag{19}$$

$$y(t) = [1 \quad 0]\begin{bmatrix} x_1(t) \\ x_2(t) \end{bmatrix}. \tag{20}$$

The corresponding discrete model is as in (16), with

$$D(\delta) = \delta^2 + \Delta\beta\delta + \beta, \tag{21}$$

$$\beta = \frac{2(1 - \cos \omega_0\Delta)}{\Delta^2}. \tag{22}$$

Combining (15), (16), and (17) leads to the following composite model:

$$\tilde{A}(\delta)y(t) = \tilde{B}(\delta)u(t) + \tilde{C}(\delta)\epsilon(t), \tag{23}$$

where

$$\tilde{A}(\delta) = G(\delta)D(\delta)A(\delta), \tag{24}$$

$$\tilde{B}(\delta) = G(\delta)D(\delta)B(\delta), \tag{25}$$

$$\tilde{C}(\delta) = F(\delta)D(\delta). \tag{26}$$

In this case of stochastic disturbances, the model (23)-(26) can be derived directly using Kalman filter results for systems having purely deterministic disturbances [1,13,36].

Note that in the presence of disturbances it is not valid, in general, to assume that $\tilde{A}$, $\tilde{B}$, and $\tilde{C}$ are relative prime. Also, to ensure that subsequent algorithms are "differentiator" free, it is helpful to constrain G and F such that degree F - degree G = degree A. We thus write F = JH, where degree H = degree G and degree J = degree A.

In the following we will use ρ as a general operator denoting respectively δ (discrete time) and d/dt (continuous time).

Where no confusion is possible, we will omit the operator ρ from polynomials and the time argument t from functions of time.

We will also use the following notation to unify the continuous and discrete case:

(a) $\mathscr{S}_0^t$ will denote

$$\int_0^t d\tau \quad \text{in continuous time}$$

$$\Delta \sum_0^t \quad \text{in discrete time}$$

(b) $\mathscr{L}_2$ will denote the spaces

$\mathscr{L}_2$ in continuous time

ℓ_2 in discrete time.

In the next section we address the question of the estimation of the parameters in the model (23).

III. PARAMETER ESTIMATION WITH FIXED NOISE FILTERS

There are two aspects to the estimation of the parameters in (23): estimation of the system transfer function, that is, A, B; and estimation of the noise model, that is, F, D, G.

We will begin by assuming that the polynomials F, D, G are approximately known. Later we shall indicate how these parameters can also be estimated. The model (23) can be written as

$$AD\left(\frac{G}{F}y\right) = BD\left(\frac{G}{F}u\right) + D\epsilon. \tag{27}$$

Since $D(\rho)$ has roots on the stability boundary, special precautions are necessary to deal with the term $D\epsilon$ in (27). Let

D_s be a stable monic polynomial "close" to D. Two discrete-time examples of such a polynomial are given below.

(i) If d(t) represents a constant disturbance, then $D = \delta$ and D_s can be taken as $\delta + h$ for small h.

(ii) If d(t) is a sine wave, then $D(\delta)$ is as in (21) and we can take

$$D_s(\delta) = \delta^2 + \Delta\gamma\beta\delta + \beta, \tag{28}$$

where

$\gamma > 1$ (say, 1.1).

Operating on (27) by $1/D_s$ gives

$$Ay_f = Bu_f + \epsilon_f, \tag{29}$$

where

$$y_f = \frac{GD}{FD_s} y = \frac{GD}{JHD_s} y, \tag{30}$$

$$u_f = \frac{GD}{FD_s} u = \frac{GD}{JHD_s} u, \tag{31}$$

$$\epsilon_f = \frac{D}{D_s}\epsilon \simeq \epsilon. \tag{32}$$

Equation (29) can be written in regression form as

$$\bar{y}_f = \phi^T\theta_0 + \epsilon_f, \tag{33}$$

where

$$\bar{y}_f = \frac{GD}{HD_s} y,$$

$$\phi^T = \left[y_f, \ldots, \rho^{n-1}y_f, u_f, \ldots, \rho^{n-1}u_f\right], \tag{34}$$

$$\theta_0^T = \left[j_0 - a_0, \ldots, j_{n-1} - a_{n-1}, b_0, \ldots, b_{n-1}\right]. \tag{35}$$

From (33) we can define the prediction error $e(\theta)$ as

$$e(\theta) = \bar{y}_f - \hat{y}_f(\theta),$$

where

$$\hat{y}_f(\theta) = \phi^T\theta.$$

We wish to adjust θ so as to minimize the prediction error. This immediately leads to the following least squares algorithm:

$$\rho\hat{\theta} = \frac{\alpha P\phi e(\hat{\theta})}{1 + \phi^T P\phi + \phi^T\phi}, \tag{36}$$

$$\rho P = \frac{-\alpha P\phi\phi^T P}{1 + \phi^T P\phi + \phi^T\phi}, \tag{37}$$

where

$$0 < \alpha \leq 1/\Delta, \tag{38}$$

$$e(\hat{\theta}) \triangleq \bar{y}_f - \phi^T\hat{\theta}. \tag{39}$$

It is desirable to slightly modify this algorithm for practical use. Two standard modifications are as follows.

(i) Incorporation of a dead zone in the prediction error $e = \bar{y}_f - \phi^T\hat{\theta}$. This modification allows bounded noise [1] to be accommodated and is also helpful in mitigating the effects of unmodeled dynamics [15,37].

(ii) The gain of the parameter estimator should be prevented from going to zero and this is achieved by adjusting P. Many methods exist for doing this, for example, exponential data weighting [achieved by multiplying the right-hand side of (37) by $1/\lambda$, $0 < \lambda < 1$]; covariance modification [achieved by adding $Q > 0$ to the right-hand side of (37]; covariance resetting [achieved by resetting P to $(1/k_0)I$ based on some criterion].

There are also variants of these basic strategies including variable forgetting factor algorithms [16], constant trace algorithms [17], etc.

A special case of (36) is the gradient algorithm where $P = (1/k_0)I$ for all time. This algorithm is very simple but generally exhibits slow convergence.

For the purposes of analysis we shall assume a modified least squares algorithm having the following properties:

Assumption A

(1) P increases when modified;

(2) $\lambda_{max}(P)$ has an upper bound;

These conditions can be readily achieved, for example, by use of the following algorithm.

(a) Initialize P with $(C_1/m)I$.

(b) Replace (37) by

$$\rho P = \frac{-\alpha\left[P\phi\phi^T P - \frac{P}{C_1}(\phi^T P^2 \phi)\right]}{1 + \phi^T P \phi + \phi^T \phi}.$$

We then have the following theorem.

Theorem 1

The algorithm (37) modified as above gives

$\lambda_{max} P \leq \text{trace } P = C_1$;

P increases when modified.

Proof. Straightforward since trace $(\rho P) = 0$. See [38] for details. □

This algorithm gives constant estimator gain (constant trace for P).

In the subsequent convergence analysis we will assume that the noise term ϵ_f is zero. This will allow us to develop the essential ideas without complicating the analysis. Results are available in the literature covering other cases, for example,

ϵ_f bounded [14], ϵ_f a stochastic colored noise process [1,18] and ϵ_f denoting the response due to unmodeled dynamics [37,39].

For future use we define the following quantities using the error (39):

normalized error: $\tilde{e} \triangleq e/(1 + \phi^T\phi)^{1/2}$ (40)

a posteriori error: $\bar{e} \triangleq e/(1 + \phi^T\phi)$ (41)

The following two lemmas will be useful in the convergence analysis. As before, the discrete-time results are obtained with $\rho = \delta$ and the continuous-time results are obtained with $\rho = d/dt$ and $\Delta = 0$.

Lemma 1: Matrix Inversion Lemma

If P is defined as in (37), then

$$\rho P^{-1} = \frac{\alpha\phi\phi^T}{1 + (1 - \alpha\Delta)\phi^T P\phi + \phi^T\phi}, \tag{42}$$

where P is any symmetric positive-definite $n \times n$ matrix, ϕ is any n vector, both P and ϕ are a function of time t, which has been omitted for brevity, and $0 < \alpha \leq 1/\Delta$.

Proof. The result follows from the discrete matrix inversion lemma by substituting $q = \delta\Delta + 1$. The proof for $\rho = d/dt$ follows directly from the product rule for differentiation applied to $I = PP^{-1}$. □

Lemma 2: Operator Product Lemma

$$\rho(x^T Ax) = [(1 + \rho\Delta)x^T][(1 + \rho\Delta)(A + A^T)][\rho x] + x^T[\rho A]x, \tag{43}$$

where A is any $n \times n$ matrix, x is an n vector, and $\Delta = 0$ if $\rho = d/dt$.

Proof. Straightforward using the definition of ρ. □

We then have the following result.

Theorem 2

If the parameter estimator (36)-(39) (modified to achieve Assumption A) is applied to the system (29) with $\epsilon_f = 0$, then

(a) $\hat{\theta}$, $\rho\hat{\theta}$, and $\tilde{e}$ are uniformly bounded;

(b) $\tilde{e}$ and $\rho\hat{\theta}$ belong to $\mathscr{L}_2$.

Proof. First consider the least squares algorithm without modification. We introduce the nonnegative time function defined by

$$V \triangleq \frac{1}{2}\tilde{\theta}^T P^{-1}\tilde{\theta}, \tag{44}$$

where

$$\tilde{\theta} = \hat{\theta} - \theta_0. \tag{45}$$

Then, from the operator product lemma and the matrix inversion lemma,

$$\rho V = \frac{-\alpha e^2}{1 + \phi^T P\phi + \phi^T\phi} \leq 0, \tag{46}$$

since $0 < \alpha \leq 1/\Delta$.

Thus V is monotone, nonincreasing, bounded from below, and therefore converges.

Since $P(t)^{-1}$ is monotone increasing, it follows that

$$\frac{1}{2}\,\tilde{\theta}(t)^T W\tilde{\theta}(t) \leq V(t)$$

for all $t \geq 0$, where $W \triangleq P(0)^{-1}$ is positive definite. Hence $||\tilde{\theta}(t)||_W$ is nonincreasing, so that $\tilde{\theta}$ and $\hat{\theta}$ are uniformly bounded. This in turn implies $\tilde{e}$ is uniformly bounded.

From (46), we have

$$\rho V \leq -\alpha\frac{e^2}{1 + \phi^T P\phi + \phi^T\phi}.$$

Hence $e^2/(1 + \phi^T P\phi + \phi^T\phi) \in \mathscr{L}_2$ and thus, since P is monotone nonincreasing,

$$\frac{e^2}{1 + [1 + \lambda_{max} P(0)]\phi^T\phi} \in \mathscr{L}_2.$$

This establishes $\tilde{e} \in \mathscr{L}_2$ and using (36) it follows that $\rho\hat{\theta} \in \mathscr{L}_2$.

The modification to P ensures that P is increased when adjusted, hence P^{-1} decreases when adjusted and thus $\frac{1}{2}\tilde{\theta}^T P^{-1}\tilde{\theta} \leq \frac{1}{2}\tilde{\theta}(\overline{P})^{-1}\tilde{\theta}$, where $\overline{P}$ is the value of P prior to modification. Hence the above analysis applies to the modified P with a less than or equal to sign in (46).

$\rho\hat{\theta}$ can be seen to be uniformly bounded as follows:

$$(\rho\hat{\theta})^T\rho\hat{\theta} = \alpha^2 \frac{\tilde{\theta}^T\phi(\phi^T P^2\phi)\phi^T\tilde{\theta}}{(1 + \phi^T P\phi + \phi^T\phi)^2} \leq \alpha^2 \frac{\lambda_{max} P(\phi^T\tilde{\theta})^2}{1 + \phi^T\phi}$$

$$\leq \alpha^2\left(\lambda_{max} P\, \tilde{e}^2\right). \tag{47}$$

The result follows from the fact that both $\tilde{\theta}$ and $\lambda_{max}(P)$ are uniformly bounded. □

Remark 1

Note that the above theorem has been derived under very weak assumptions. Thus it is not necessarily true that $\hat{\theta}$ converges to the true value of θ. However, the results, as presented, are sufficient to establish convergence of a wide class of adaptive control laws. If it is required that $\hat{\theta}$ converge to the true value, then additional assumptions are required. In particular, it suffices to have the input persistently exciting. Details of how this can be achieved are given in [1].

IV. PARAMETER ESTIMATION WITH UNKNOWN NOISE FILTERS

When the noise filters are unknown, the prediction error can still be computed. However $e(\theta)$ is now a nonlinear function of θ:

$$e(\theta) = \frac{GD}{FD_s}\{Ay - Bu\}, \tag{48}$$

where G, D, F, D_s, A, and B all depend on θ.

Given an estimate $\hat{\theta}$ of θ, (48) can be linearized as follows:

$$e(\theta) \simeq e(\hat{\theta}) + \frac{\partial e}{\partial \theta}(\theta - \hat{\theta}) = \left(e(\hat{\theta}) - \frac{\partial e}{\partial \theta}\right) - \xi^T\theta, \tag{49}$$

where

$$\xi^T \triangleq -\frac{\partial e}{\partial \theta}\bigg|_{\theta=\hat{\theta}}. \tag{50}$$

The linearized error (50) suggests the following nonlinear least squares algorithm:

$$\rho\hat{\theta} = \frac{\alpha P\xi e(\hat{\theta})}{1 + \xi^T P\xi}, \tag{51}$$

$$\rho P = \frac{-\alpha P\xi\xi^T P}{1 + \xi^T P\xi}, \tag{52}$$

where

$$0 < \alpha \leq 1/\Delta, \tag{53}$$

$$e(\hat{\theta}) = \frac{GD}{FD_s}\{Ay - Bu\}\bigg|_{\theta=\hat{\theta}} \tag{54}$$

The above algorithm has been well studied in the stochastic case (see [19] in particular). Properties similar to those found for the fixed noise filter case can also be derived (see [1,19]). In practice, some form of projection is usually necessary to ensure that G, F remain stable. Also, since D/D_s is basically a notch filter, it is necessary to make the notch wide when the parameter estimates are poor since otherwise there is little gradient information available with which to improve the estimates.

In the next section we will describe a possible control law design method for the model (27).

V. STOCHASTIC CONTROL

Given the model (23), there exists a large number of possible control system design methods. We describe below a particular algorithm based on pole assignment.

We assume that the set point y^* belongs to the class of signals that can be modeled as

$$S(\rho)y^* = 0; \tag{55}$$

For example, if y^* = constant, then $S(\rho) = \rho$.

We also factor B into the form $B_\alpha B_\beta$, where B_α is stable and well damped. We will cancel B_α in the design procedure.

The pole-assignment control law is then of the form

$$B_\alpha LDGSu = -Py + \eta, \tag{56}$$

where η is an external input and where L and P are found by solving the following pole-assignment equation:

$$C_s A^* = LDGSA + B_\beta P, \tag{57}$$

where $C_s A^*$ is stable and C_s is chosen as a stable approximation to $S\tilde{C} = FDS$.

Important features of the above design are as follows.

(a) The control law has S and D in the denominator. Thus, with $\eta = P\,y^*$, a form of the internal model principle [20] has been used. In the noise-free case, this gives perfect tracking of the deterministic set point and elimination of the deterministic disturbance [20].

(b) With noise and $\eta = Py^*$, it is readily seen that the closed-loop system satisfies

$$C_s A^* \left[y - y^* - \frac{LS\tilde{C}}{A^* C_s} \epsilon \right] = 0. \quad (58)$$

Thus the steady-state tracking error is

$$\xi = \left(LS\tilde{C}/A^* C_s \right) \epsilon. \quad (59)$$

If C_s is close to $S\tilde{C}$, then the principal term in (59) arises from the filter L/A^*. This is a result of the shifting of B_β into the stable region.

(c) If $B_\beta = 1$, then all of the system zeros are canceled. In this case, (57) can be trivially solved by back substitution.

(d) If $\eta = Py^*$, then unity feedback is used. It is well known in classical control that this renders the closed-loop performance insensitive to changes in the system or control law. In fact, the well-known PID controller is simply a special case of (56) in which $\tilde{B}_\alpha = L = D = G = 1$, $S = \rho$, and P has order 2.

(e) The above design method includes, as special cases, many other procedures, for example, model reference control, minimum variance control, etc. To illustrate, we consider deterministic model reference control. The system (27) is then specialized to the case $\epsilon = 0$, $F = D = G = 1$. The polynomials A, B will be assumed to be of order n, m, respectively, and the reference model will be taken to have input r and output y^*. Equation (55) no longer applies to y^*; instead y^* and r are assumed to be realted by the reference model

$$Ey^* = \bar{H}r, \quad (60)$$

where E, $\bar{H}$ have order $2n - m$ and n, respectively.

We require the plant to be minimum phase and put $B = B_\alpha$, $B_\beta = 1$, and $C_s A^* = E$. The feedback law achieving model reference control is then as in (56), with DGS = 1, $\eta = \bar{H}r$, giving $Ey = \bar{H}r$ in steady state. Equation (57) reduces to

$$LA + P = E. \tag{61}$$

This equation is called the "prediction equality" in the discrete case [1] and is trivial to solve by back substitution. Further discussion of the continuous case is contained in [21].

(f) An interesting question concerning the zeros of the polynomial B arises in relation to discrete model reference control of continuous-time systems having relative degree greater than or equal to two. We consider a continuous-time system having transfer function $B(s)/A(s)$, where $A(s) = s^n + a_1 s^{n-1} + \cdots + a_n$, $B(s) = b_0 s^m + b_1 s^{m-1} + \cdots + b_m$, $b_0 \neq 0$. The corresponding discrete transfer function (in terms of the shift operator) is denoted $\bar{B}(q)/\bar{A}(q)$.

It is well known that if $p_1, \ldots, p_n$ denotes the zeros of $A(s)$, then the zeros of $\bar{A}(q)$ are $e^{p_1\Delta}, \ldots, e^{p_n\Delta}$. However, the situation for the zeros of $\bar{B}(q)$ is more complex. Astrom, Hagander, and Sternby [35] have shown that for rapid sampling, if $B(s)$ has zeros $z_1, \ldots, z_m$, then $\bar{B}(z)$ can be expressed as $\bar{B}(z) = \bar{B}_1(z)\bar{B}_2(z)$ where $B_1(z)$ has degree m and has zeros converging, as $\Delta \to 0$, to $e^{z_1\Delta}, \ldots, e^{z_m\Delta}$ and $B_2(z)$ has degree $n - m - 1$ and converges, as $\Delta \to 0$, to a unique polynomial form $\bar{\bar{B}}_2(z)$ for each value of $n - m - 1$. For $n - m \geq 2$, the composite polynomial $\bar{B}(z)$ has zeros outside (or on) the unit circle and its coefficients are complicated functions of the zeros of $B(s)$. These two facts seem to imply that discrete-time model reference control, which involves cancellation of all the plant zeros, cannot strictly be applied to continuous systems having relative degree greater than or equal to two. On the other hand, continuous-time model reference control of continuous systems having relative degree greater than one is straightforward. This paradox is resolved using the delta operator. The appropriate delta model is

$$A'(\delta)y(K\Delta) = B'(\delta)u(k\Delta), \tag{62}$$

where

$$\begin{aligned} A'(\delta) &= \delta^n + a_1'\delta^{n-1} + \cdots + a_n', \\ B'(\delta) &= B_\epsilon'(\delta) + B_R'(\delta); \\ B_\epsilon'(\delta) &= \epsilon_1\delta^{n-1} + \cdots + \epsilon_{n-m-1}\delta^{m+1}, \\ B_R'(\delta) &= b_0'\delta^m + \cdots + b_m' \end{aligned} \tag{63}$$

An expression for $B'(\delta)$ and $A'(\delta)$ can be derived as follows: expanding $T(s)/s = B(s)/sA(s)$ using partial fractions, we obtain

$$\frac{T(s)}{s} = \sum_{i=1}^{N}\left\{\frac{a_{i0}}{s} + \sum_{j=1}^{\nu_i}\frac{a_{ij}}{(s+p_i)^j}\right\}, \tag{64}$$

where $\Sigma_{i=1}^{N}\nu_i = n$. Hence the z-domain transfer function is

$$\begin{aligned} \overline{T}(z) &= (1 - z^{-1})\sum_{i=1}^{N}\left\{\frac{a_{i0}}{1 - z^{-1}} + \sum_{j=1}^{\nu_i}\frac{a_{ij}}{(j-1)!}\right. \\ &\qquad \left.(-1)^{n-1}\frac{\partial^{(j-1)}}{\partial p_i^{(j-1)}}\left[\frac{z}{z - e^{-p_i\Delta}}\right]\right\} \\ &= \sum_{i=1}^{N}\left\{a_{i0} + \sum_{j=1}^{\nu_i}\frac{(z-1)a_{ij}(-1)^{j-1}}{(j-1)!}\right. \\ &\qquad \left.\times\frac{\partial^{(j-1)}}{\partial p_i^{(j-1)}}\left[\frac{1}{z - e^{-p_i\Delta}}\right]\right\}. \end{aligned}$$

Now the delta transfer function is $T'(\delta) = \overline{T}(z)$ evaluated at

$$z = 1 + \delta\Delta$$

$$= \sum_{i=1}^{N} \left\{ a_{i0} + \sum_{j=1}^{\nu_i} \delta a_{ij} \frac{(-1)^{j-1}}{(j-1)!} \frac{\partial^{(j-1)}}{\partial p_i^{(j-1)}} \times \left[\frac{1}{\delta + (1 - e^{-p_i\Delta})/\Delta} \right] \right\}.$$

This immediately gives expressions for $B'(\delta)$ and $A'(\delta)$. Note that $A'(\delta)$, $B'(\delta)$ have degree n and n - 1, respectively.

In the limit as $\Delta \to 0$, we have

$$\lim_{\Delta\to 0} T'(\delta) = \sum_{i=1}^{n} \left\{ a_{i0} + \sum_{j=1}^{\nu_i} \delta a_{ij} \frac{(-1)^{j-1}}{(j-1)!} \frac{\partial^{(j-1)}}{\partial p_i^{(j-1)}} \left[\frac{1}{\delta + p_i} \right] \right\}$$

$$= \sum_{i=1}^{n} \left\{ a_{i0} + \sum_{j=1}^{\nu_i} \frac{\delta a_{ij}}{(\delta + p_i)^j} \right\}$$

$$= T(s) \quad \text{evaluated at} \quad s = \delta.$$

We thus see that in (64) as $\Delta \to 0$, $a_i' \to a_i$ $(i = 1, \ldots, n)$; $b_i' \to b_i$ $(i = 1, \ldots, m)$; and $\epsilon_i \to 0$ $(i = 1, \ldots, n - m - 1)$; where $\{a_i\}$, $\{b_i\}$ are the parameters of the continuous model.

Therefore, with rapid sampling, the delta model allows the "extraneous" zeros arising from the sampling process to be isolated. Furthermore, it is shown in [22] that if one ignores the polynomial $B_\epsilon'(\delta)$ in designing a discrete-time model reference controller, then provided the continuous-time system is minimum phase, there exists a $\Delta_1 > 0$ such that with $\Delta \leq \Delta_1$ the resulting closed-loop system is guaranteed stable *irrespective of the relative degree* of the continuous-time system. This allows discrete model reference control to be implemented without difficulty on continuous systems of arbitrary relative degree.

VI. ADAPTIVE CONTROL

In principle, to design an adaptive control law, one only need combine a parameter estimation algorithm from Section III or IV with a control law synthesis procedure from Section V. There are clearly many possible algorithms. In the absence of physical reasons to act otherwise, a reasonable choice would seem to be to use the least squares algorithm modified to satisfy Assumption A together with the robust stochastic control law given in (56) and (57).

An issue that has been addressed extensively in the literature on adaptive control is the question of global convergence of the algorithms under ideal assumptions. A convergence theory is important since one would have little confidence in an algorithm that could not be shown to converge under ideal conditions. Also, the theory gives insight into practical implementation issues since factors which arise in the theory are often of practical importance. We therefore give below a general convergence tool that can be used to establish global convergence of a wide class of adaptive control algorithms. We will later apply this theory to a particular algorithm.

Again, we will treat the continuous and discrete cases side by side. However, the continuous case is technically more difficult and thus we will retain sufficient generality to cover this case. Global convergence proofs for the continuous-time case were first presented in [23,24,25], where it is shown that convergence depends on a growth condition on the regresssion vector ϕ and its derivations up to order $n - m - 1$. We note that the case $n - m = 1$ is particularly simple because no derivations of ϕ are involved. This partially explains why the case $n - m = 1$ was resolved much earlier in the literature than the general case.

Let $H(t) \in R^{(n-m)*\dim(\phi)}$ be defined as follows:

$$H(t) = \begin{bmatrix} \phi(t) \\ \rho\phi(t) \\ \vdots \\ \rho^{n-m-1}\phi(t) \end{bmatrix}. \tag{65}$$

(In discrete time, ρ simple shifts ϕ and it turns out that $\rho\phi$, etc., need not be considered in the analysis.)

The following lemma links together the properties of the parameter estimation and control calculations in order that convergence can be established.

Lemma 3: Stability Lemma

Suppose the estimator is such that the following hypothesis is satisfied:

$$\text{E1:} \quad \tilde{e} \in \mathscr{L}_2. \tag{66}$$

Suppose the controller is such that the following growth hypothesis is satisfied:

C1: There exists a constant $d \in (0, \infty)$ such that

$$\|H(t)\|^2 \leq d + d \int_0^t \|H(\tau)\|^2 \tilde{e}(\tau)^2. \tag{67}$$

Then H is uniformly bounded.

Proof. From C1 and Gronwall's lemma (see, for example, [26, p. 252], we have

$$\|H(t)\|^2 \leq d \exp\left\{\int_0^t \tilde{e}(\tau)^2\right\}. \tag{68}$$

In continuous time, this is a direct consequence of Gronwall's lemma, and in discrete time, Gronwall's lemma gives

$$\|H(t)\|^2 \leq d \prod_{0\leq\tau\leq t} [1 + \tilde{e}(\tau)^2].$$

Hence

$$\ln\|H(t)\|^2 \leq \ln d + \sum_{0}^{t} \ln[1 + \tilde{e}(\tau)^2]$$

$$\leq \ln d + \sum_{0}^{t} \tilde{e}(\tau)^2.$$

Thus

$$\|H(t)\|^2 \leq \exp\left\{\ln d + \sum_{0}^{t} \tilde{e}(\tau)^2\right\}.$$

The result follows from (68) since $\tilde{e} \in \mathscr{L}_2$. □

Remark 2

The above lemma covers the noise-free continuous and discrete results. Actually, in the discrete case, it has been traditional [27] to replace (67) by the stronger condition

$$\|\phi(t)\|^2 \leq C_1 + C_2 \max_{0\leq\tau\leq t} e(\tau)^2. \tag{69}$$

To see that this implies (67) we argue as follows: from (69),

$$\|\phi(t)\|^2 \leq C_1 + C_2 \max_{0<\tau<t} \{1 + \|\phi(\tau)\|^2\}\tilde{e}(\tau)^2$$

$$\leq C_1 + C_2 \sum_{0}^{t} \{1 + \|\phi(\tau)\|^2\}\tilde{e}(\tau)^2$$

$$\leq C_1' + C_2 \sum_{0}^{t} \|\phi(\tau)\|^2 \tilde{e}(\tau)^2,$$

using the fact that $\tilde{e} \in \mathscr{L}_2$. □

Remark 3

In the discrete stochastic case [1,18], the growth condition (69) is of the form

$$\text{C1}': \quad \frac{1}{t}\sum_{0}^{t} \|\phi\|^2 \leq K_1 + \frac{K_2}{t}\sum_{0}^{t}(e - v)^2 \quad \text{(a.s.)},$$

where v(t) is a noise sequence satisfying the following conditions for some increasing sequence of σ algebras $\mathcal{F}_{t-1}$:

$$(1) \quad E\{v(t)\,|\mathcal{F}_{t-1}\} = 0 \qquad \text{(a.s.)};$$

$$(2) \quad E\left\{v(t)^2\,|\mathcal{F}_{t-1}\right\} = \gamma^2 \qquad \text{(a.s.)};$$

$$(3) \quad \limsup_{N\to\infty} \frac{1}{N}\sum_{0}^{N} v(t)^2 < \infty \qquad \text{(a.s.)}.$$

Also, the properties of parameter estimator E1 have the following form:

$$\text{E1}': \quad \sum_{0}^{\infty} \frac{(e - v)^2}{r} < \infty \qquad \text{(a.s.)},$$

where

$$r = 1 + \sum_{0}^{t} \|\phi\|^2.$$

The stochastic equivalent to Lemma 3 then states that E1' and C1' imply

$$\limsup_{N\to\infty} \frac{1}{N}\sum_{0}^{N} \|\phi\|^2 < \infty \qquad \text{(a.s.)},$$

which is equivalent to (67).

The close connection between the stochastic results and the noise-free result in Lemma 3 is thus clear. □

The above result can be used to establish global convergence for a very wide class of adaptive control algorithms (see, for example, [24,25,27]). The principle difficulty is to establish condition C1.

In the analysis of many algorithms we need to have a bound on the "swapping" error caused when a term like $(1/E_2)(\phi^T\tilde{\theta})$ is replaced by $[(1/E_2)\phi]^T\tilde{\theta}$. That this error is small when $\tilde{\theta}$ changes slowly is intuitively obvious and is made precise in the following result. First, we note that $\psi = (1/E_2)\phi$ satisfies the following set of equations:

$$\rho H = AH + b\phi^T, \qquad \psi = c^T H,$$

where A is in controller canonical form, the characteristic polynomial of A is $E_2(s)$, $b = e_{n-m-1}$, and $c = e_1$. We note that

$$\left(\frac{1}{E_2}\,\phi\right)^T\tilde{\theta} = \psi^T\tilde{\theta} = c^T H\tilde{\theta},$$

and that $\xi = (1/E_2)(\phi^T\tilde{\theta})$ is the solution of

$$\rho g = Ag + b(\phi^T\tilde{\theta}), \qquad \xi = c^T g.$$

Hence we can compute $H\tilde{\theta} - g$ and this automatically yields

$$\left(\frac{1}{E_2}\,\phi\right)^T\tilde{\theta} - \frac{1}{E_2}(\phi^T\tilde{\theta}) = c^T[H\tilde{\theta} - g].$$

Lemma 4 (24): The Swapping Lemma

(a) $H\tilde{\theta} - g = \Phi_A(t) * H\dot{\tilde{\theta}}$,

where $\Phi_A(t)$ is the state transition matrix for the system $\rho x = Ax$ and where we have used $\dot{\tilde{\theta}}$ to denote $\rho\tilde{\theta}$.

(b) $\left(\frac{1}{E_2}\,\phi\right)^T\tilde{\theta} - \frac{1}{E_2}(\phi^T\tilde{\theta}) = c^T[H\tilde{\theta} - g] = c^T\Phi_A(t) * H\dot{\tilde{\theta}}$.

Proof.

$$\rho H\tilde{\theta} = AH\tilde{\theta} + b\phi^T\tilde{\theta} + H\dot{\tilde{\theta}}, \qquad \rho g = Ag + b\phi^T\tilde{\theta}.$$

Hence

$$\rho[H\tilde{\theta} - g] = A[H\tilde{\theta} - g] + H\dot{\tilde{\theta}}. \quad \square$$

Since A is stable, the desired result follows.

One further result is needed.

Lemma 5

Suppose the weighting function W is exponentially stable. Then

(a) there exist constants C_1, $C_2 < \infty$ such that

$$\|(W * u)(t)\| \leq C_1 + C_2 \sup_{0 \leq \tau \leq t} \|u(\tau)\|;$$

(b) if W is also strictly proper and u is piecewise continuous, then there exists a $d < \infty$ such that

$$\|(W * u)(t)\|^2 \leq d + d \int_0^t \|u(s)\|^2;$$

(c) if W is also strictly proper, u_1 is uniformly bounded, and u_2 is piecewise continuous, then there exists a $d < \infty$ such that

$$\|W * (u_1 + u_2)(t)\|^2 \leq d + d \int_0^t \|u_2(s)\|^2.$$

Proof. Parts (a) and (b) are straightforward (see, for example, [1, Appendix B]). Part (c) follows from the trivial observation that

$$\|W * (u_1 + u_2)\|^2 = \|(W * u_1) + (W * u_2)\|^2$$
$$\leq 2\|W * u_1\|^2 + 2\|W * u_2\|^2. \quad \square$$

We will illustrate the application of Lemma 3 using a very simple algorithm. This will show the basic proof technique. A very general result of the same type is proved in [28].

We consider a simple case of the model (27) in which D = 1, G = 1, F = J (of order n), and b_m is known. We also treat only the case $n - m = 1$ with $\epsilon_f = 0$; (33) becomes

$$y = \psi^T\theta_0 + b_m\rho^m u_f, \tag{70}$$

where

$$\psi^T = \left[y_f, \ldots, \rho^{n-1}y_f, u_f, \ldots, \rho^{n-2}u_f\right], \quad (71)$$

$$\theta_0 = [j_0 - a_0, \ldots, j_{n-1} - a_{n-1}, b_0, \ldots, b_{n-2}], \quad (72)$$

$$y_f = (1/J)y, \quad u_f = (1/J)u. \quad (73)$$

We use the parameter estimator (36)-(38) (modified to achieve Assumption A), where ϕ is replaced by ψ and $e(\hat{\theta})$ now has the form

$$e(\hat{\theta}) = y - b_m\rho^m u_f - \psi^T\hat{\theta}. \quad (74)$$

We use the certainty equivalence form of the model reference control law, that is, given $\hat{\theta}$, we solve the following equation for L, P (order 1, n - 1, respectively):

$$\hat{L}\hat{A} + \hat{P} = E, \quad \text{with reference model} \quad Ey^* = \overline{H}r. \quad (75)$$

where $\hat{L}\hat{A}$ denotes the product of the operators $\hat{L}$ and $\hat{A}$. We implement the control law as

$$LBu_f = -Py_f + Hr_f, \quad r_f = (1/J)r. \quad (76)$$

We then have the following lemma.

Lemma 6

Provided the system is minimum phase, then the above algorithm ensures that Condition C1 is satisfied.

Proof. Equation (74) can be rewritten as

$$e = \hat{A}y_f - \hat{B}u_f, \quad (77)$$

where $\hat{b}_m = b_m$.

Operating on the left-hand side of (77) by $\hat{L}$ gives

$$\hat{L}e = \hat{L} \cdot \hat{A}y_f - \hat{L} \cdot \hat{B}u_f, \quad (78)$$

where we have used the notation

$$\hat{L} \cdot \hat{A}y_f = \hat{L}\{\hat{A}\{y_f\}\}. \quad (79)$$

Note that, in general, $\hat{L} \cdot \hat{A}y_f \neq \hat{L}\hat{A}y_f \neq \hat{A} \cdot \hat{L}y_f$ due to the time-varying nature of $\hat{A}$ and $\hat{L}$.

Equation (78) can be rewritten as

$$\hat{L}\hat{A}y_f + \{\hat{L} \cdot \hat{A} - \hat{L}\hat{A}\}y_f - \hat{L}\hat{B}u_f - \{\hat{L} \cdot \hat{B} - \hat{L}\hat{B}\}u_f = \hat{L}e. \tag{80}$$

Using (76) and (75) we have

$$Ey_f + \{\hat{L} \cdot \hat{A} - \hat{L}\hat{A}\}y_f - \{\hat{L} \cdot \hat{B} - \hat{L}\hat{B}\}u_f = \hat{L}e + \bar{H}r_f. \tag{81}$$

From (75), we have $\hat{L} = \rho + \hat{\ell}_0$, and thus

$$(\hat{L} \cdot \hat{A} - \hat{L}\hat{A}) = \dot{\hat{A}}, \tag{82}$$

$$(\hat{L} \cdot \hat{B} - \hat{L}\hat{B}) = \dot{\hat{B}}, \tag{83}$$

where

$$\dot{\hat{A}} = \dot{\hat{a}}_{n-1}\rho^{n-1} + \cdots + \dot{\hat{a}}_0, \tag{84}$$

$$\dot{\hat{B}} = \dot{\hat{b}}_{n-2}\rho^{n-2} + \cdots + \dot{\hat{b}}_0, \tag{85}$$

and

$$\dot{\hat{a}}_{n-1} = \left\{\rho\hat{a}_{n-1}\right\}, \quad \text{etc.}, \tag{86}$$

$$\left[-\dot{\hat{a}}_0, \ldots, -\dot{\hat{a}}_{n-1}, \dot{\hat{b}}_0, \ldots, \dot{\hat{b}}_{n-2}\right]^T = \dot{\hat{\theta}} = \dot{\tilde{\theta}}. \tag{87}$$

Substituting (84), (85) into (81) gives

$$Ey_f = \hat{L}e + \bar{H}r_f + \psi\,\dot{\tilde{\theta}} = \left(\rho + \hat{\ell}_0\right)e + \bar{H}r_f + \psi^T\dot{\tilde{\theta}}. \tag{88}$$

Now using the fact that $\hat{\ell}_0$ is uniformly bounded (since $\hat{\theta}$ is uniformly bounded), we have from Lemma 5 that there exists $0 \leq d_{1i}, d_{2i} < \infty$ such that

$$|\rho^i y_f(t)|^2 \leq d_{1i} + d_{2i} \mathscr{S}_0^t \{e^2 + ||\psi||^2 ||\dot{\tilde{\theta}}||^2\}$$
$$i = 0, \ldots, n-1. \tag{89}$$

Now from (47),

$$||\dot{\tilde{\theta}}||^2 \leq \alpha^2(\lambda_{max}P)\tilde{e}^2.$$

Then using Assumption A, there exists $0 < C < \infty$ such that

$$\|\dot{\tilde{\theta}}\|^2 \le C\tilde{e}^2. \tag{90}$$

Substituting into (89) gives

$$|\rho^i y_f(t)|^2 \le d'_{1i} + d'_{2i} \mathscr{S}_0^t \{(1 + \|\psi\|^2)\tilde{e}^2 + c\|\psi\|\tilde{e}^2\}$$

$$\le d''_{1i} + d''_{2i} \mathscr{S}_0^t \|\psi\|^2\tilde{e}^2 \tag{91}$$

since $\tilde{e} \in \mathscr{L}_2$, $i = 0, \ldots, n - 1$.

Now

$$\rho^k u_f = \frac{A}{B} \rho^k y_f = \frac{\Sigma_{j=0}^n a_j \rho^{k+j} y_f}{B}. \tag{92}$$

Hence using (91) and Lemma 5, we have

$$|\rho^i u_f(t)|^2 \le \bar{d}_{1i} + \bar{d}_{2i} \mathscr{S}_0^t \|\psi\|^2\tilde{e}^2, \quad i = 0, \ldots, n - 2. \tag{93}$$

Combining (91) and (93), we have C1 for $n - m = 1$. □

Finally, we have the following theorem.

Theorem 3

The above algorithm ensures

(1) ψ, $\tilde{\theta}$, $\rho\tilde{\theta}$, e, $\tilde{e}$, $\bar{e}$, and u are uniformly bounded;

(2) $e \to 0$ as $t \to \infty$;

(3) $|y - y^*| \to 0$ as $t \to \infty$.

Proof.

1. $\tilde{\theta}$, $\rho\tilde{\theta}$, and $\tilde{e}$ are uniformly bounded from Theorem 2. ψ being uniformly bounded follows from Lemmas 3 and 6. This immediately implies e and $\bar{e}$ are uniformly bounded. Now, from (88), since $\bar{H}r_f$, e, ψ, and $\rho\tilde{\theta}$ are all uniformly bounded, we have from Lemma 5 that $\rho^i y_f(t)$ is uniformly bounded for $i = 0, \ldots, n$. Similarly, from (92) we have $\rho^i u_f(t)$ is uniformly bounded for

$i = 0, \ldots, n-1$. Hence $\rho\psi$ is uniformly bounded. Then, since $e = \psi^T\tilde{\theta}$, ρe is uniformly bounded. From (88), noting that E is of order (n + 1), $\rho^{n+1}y_f$ is uniformly bounded using Lemma 5. Then from (92), $\rho^n u_f$ is uniformly bounded. Also, since $u = Ju_f$, we see that u is uniformly bounded.

2. Since $\tilde{e} \in \mathscr{L}_2$ and ψ is uniformly bounded, we have $e \in \mathscr{L}_2$. Also, e and ρe are uniformly bounded. Hence $e \to 0$ as $t \to \infty$.

3. From (88) we have

$$(y - y^*) = \frac{\rho J}{E} e + \frac{J}{E}\left\{\hat{\ell}_0 e + \psi^T \dot{\tilde{\theta}}\right\} \tag{94}$$

Now since $e \to 0$, from (90), $\dot{\tilde{\theta}} \to 0$, and hence from (94) $y \to y^*$. □

The theory, as presented above, applies to the case of model reference control. In the case of the more general pole-assignment problem, an additional technical difficulty arises, namely, that one must ensure that (57) can be solved at all times. Otherwise, the theory is identical to that presented above. Solvability of (57) requires that $\hat{B}_\beta$ and $DGS\hat{A}$ be relatively prime. Three methods have been discussed in the literature for ensuring this condition.

(a) $\hat{B}_\beta$, $\hat{A}$ are constrained to lie within regions in which pole-zero cancellations do not occur (see, for example, [1, pp. 91 and 211; 37]).

(b) A persistently exciting external input is applied to ensure that $\hat{A}$, $\hat{B}$ converge to their true values, in the vicinity of which pole-zero cancellations do not occur by assumption (see, for example, [1, p. 217]).

(c) Modifications to the parameter estimator can be found that ensure that if a near pole-zero cancellation does occur, then other parameter estimates can be found by a finite search procedure that retain the key properties of the parameter estimator (as in Theorem 2) but that do not have a pole-zero cancellation (see [17] for details).

In the above analysis we have assumed perfect modeling. However, the theory can be extended to include plants having unmodeled dynamics (see for example, [37,39]).

VII. APPLICATION TO A SERVO SYSTEM

The model reference adaptive control ideas presented above have been implemented on an electromechanical servo system. The servo system has an open-loop transfer function

$$\frac{y(s)}{u(s)} = \frac{K}{s(1 + s\tau)}, \tag{95}$$

where y is the angle of the motor shaft, u the voltage input to the servo amplifier, and K and τ are the servo constants, which can be varied on-line by the operator. Parameter estimation was performed in indirect discrete time δ form with a sampling rate of 50 Hz. The estimation algorithm included a 0.15° dead zone and the constant trace modification was used as suggested in Section III to ensure that Assumption A was satisfied. The highest order coefficient in $\hat{B}$ was estimated and constrained to be greater than some small positive value to ensure arithmetic overflow did not occur in case the gain K was inadvertently set to zero. As suggested in Section V, $\hat{B}$, the portion of $\hat{B}$ corresponding to a zero near $\delta = -2/\Delta$ (i.e., $q = -1$), was ignored in the algorithm. If this is not done, a near unstable pole/zero cancellation occurs between the controller and the plant leading to robustness problems [22]. The reference model used was $1/(0.07\delta + 1)^3$. The system performance was excellent even in the presence of rapid variations of K and τ. Figure 1 shows the response of the system when K is suddenly increased from

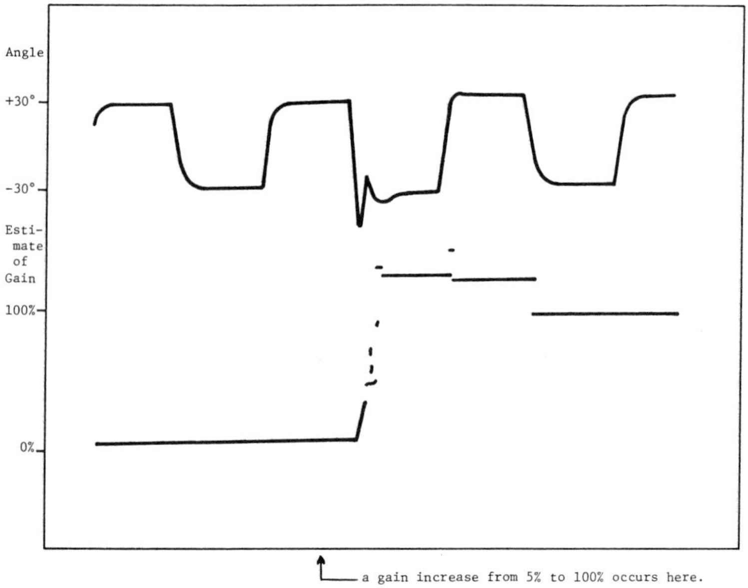

Fig. 1. Response of the servo system to an increase in K.

5 to 100%. The input to the reference model was a 1-Hz square wave of amplitude 30°. The upper trace shows that the system settles very quickly (in about 0.3 sec). The lower trace shows the estimate of the system gain.

VIII. CONCLUSION

This article has given a brief overview of a particular approach to adaptive control. Emphasis has been placed on merging the continuous and discrete cases. Also, the algorithms have been selected with robustness considerations in mind. Further information and background can be obtained by reading some of the references, especially [1] and [29-39].

REFERENCES

1. G. C. GOODWIN and K. S. SIN, "Adaptive Filtering Prediction and Control," Prentice-Hall, Englewood Cliffs, New Jersey, 1984.

2. U. BORISSON, "Self Tuning Regulators for a Class of Multivariable Systems," *Automatica 15*, No. 2, 209-217 (1979).

3. Y. BAR-SHALOM and E. TSE, "Dual Effort, Certainty Equivalence and Separation in Stochastic Control," *IEEE Trans. Autom. Control AC-19*, 494-500 (1974).

4. Y. BAR-SHALOM and E. TSE, "Concepts and Methods in Stochastic Control," *in* "Control and Dynamic Systems: Advances in Theory and Application," Vol. 12, (C. T. Leondes, ed.), Academic Press, New York, 1976.

5. J. TSCHAUNER, "Introduction à la Théorie des Systèmes Échantillonnés," p. 42, Dunod, Paris, 1963.

6. R. J. KARWOSKI, "Introduction to the Z Transform and Its Derivation," Tutorial Paper, TRW psi Products, El Segundo, California, 1979.

7. P. J. GAWTHROP, "Hybrid Self Tuning Control," *Proc. IEE 127*, 5 (1980).

8. R. C. AGARWAL and C. S. BURRUS, "New Recursive Digital filter Structures Having Very Low Sensitivity and Roundoff Noise," *IEEE Trans. Circuits Syst. CAS-22*, No. 12, 921-927 (1984).

8. R. C. AGARWAL and C. S. BURRUS, "New Recursive Digital Filter Structures Having Very Low Sensitivity and Roundoff Noise," *IEEE Trans. Circuits Syst. CAS-31*, No. 7 (1984).

9. J. M. EDMUNDS, "Identifying Sampled Data Systems Using Difference Operator Models," Control Systems Report, No. 601, University of Manchester, 1985.

10. G. ORLANDI and G. MARTINELLI, "Low Sensitivity Recursive Digital Filters Obtained via the Delay Replacement," *IEEE Trans. Circuits Syst. CAS-31*, No. 7, 654-657 (1984).

11. R. H. MIDDLETON and G. C. GOODWIN, "Improved Finite Word Length Characteristics in Digital Control Using Delta Operators," *IEEE Trans. Autom. Control AC-31*, No. 11 (1986), in press.

12. Y. KAILATH, "Linear Systems," Prentice-Hall, Englewood Cliffs, New Jersey, 1980.

13. S. W. CHAN, G. C. GOODWIN, and K. S. SIN, "Convergence Properties of the Riccati Difference Equation in Optimal Filtering of Nonstabilizable Systems," *IEEE Trans. Autom. Control AC-29*, No. 2, 110-118 (1982).

14. K. S. NARENDRA and B. B. PETERSON, "Bounded Error Adaptive Control," *Proc. IEEE Conf. Decision Control, Albuquerque, New Mexico* (1980).

15. G. C. GOODWIN, D. J. HILL, and M. PALANISWAMI, "Towards an Adaptive Robust Controller," *Proc. IFAC Symp. Identif. Syst. Parameter Estim. 7th, York, England,* 1985.

16. T. R. FORTESUE, L. S. KERSHENBAUM, and B. E. YDSTIE, "Implementation of Self-Tuning Regulators with Variable Forgetting Factors," *Automatica 17*, 831-835 (1981).

17. R. L. LEAL and G. C. GOODWIN, "A Globally Convergent Adaptive Pole Placement Algorithm without Persistency of Excitation Requirement," *IEEE Trans. Autom. Control AC-30*, No. 8, pp. 795-798 (1986).

18. G. C. GOODWIN, P. J. RAMADGE, and P. E. CAINES, "Stochastic Adaptive Control," Technical Report, Harvard University, Cambridge, Massachusetts, 1978.

19. L. LJUNG and T. SÖDERSTRÖM, "Theory and Practice of Recursive Identification," MIT Press, Cambridge, Massachusetts, 1982.

20. B. A. FRANCIS and W. M. WONHAM, "The Internal Model Principle of Control Theory," *Automatica 12*, 457-465 (1976).

21. B. EGARDT, "Unification of Some Continuous-Time Adaptive Control Schemes," *IEEE Trans. Autom. Control AC-24*, No. 4, 588-592 (1979).

22. G. C. GOODWIN, D. Q. MAYNE, and R. L. LEAL, "Rapproachment between Continuous and Discrete Model Reference Adaptive Control," *Automatica*, 199-208, March 1986.

23. A. FEUER and A. S. MORSE, "Adaptive Control of Single Input Single Output Linear System," *IEEE Trans. Autom. Control AC-23*, 557-570 (1978).

24. A. S. MORSE, "Global Stability of Parameter Adaptive Control Systems," *IEEE Trans. Autom. Control AC-25*, No. 3, 433-439 (1980).

25. K. S. NARENDRA, Y. H. LIN, and L. S. VALAVANI, "Stable Adaptive Controller Design, Part II: Proof of Stability," *IEEE Trans. Autom. Control AC-25*, 243-247 (1980).

26. C. A. DESOER and M. VIDYASAGAR, "Feedback Systems: Input and Output Properties," Academic Press, New York, 1975.

27. G. C. GOODWIN, P. J. RAMADGE, and P. E. CAINES, "Discrete Time Multivariable Adaptive Control," *IEEE Trans. Autom. Control AC-25*, 449-456 (1980).

28. G. C. GOODWIN and D. Q. MAYNE, "A Parameter Estimation Perspective of Continuous Time Adaptive Control," *Automatica*, November 1986.

29. K. J. ASTRÖM, "Theory and Applications of Adaptive Control—A Survey," *Automatica 19*, No. 5, 471-487, September 1983.

30. K. S. NARENDRA and R. V. MONOPOLI (ed.), "Applications of Adaptive Control," Academic Press, New York, 1980.

31. C. J. HARRIS and S. A. BILLINGS (ed.), "Self-Tuning and Adaptive Control:— Theory and Applications," Peregrins, London, 1981.

32. Y. Z. TSYPKIN, "Foundations of the Theory of Learning Systems," Academic Press, New York, 1973.

33. H. UNBEHAUN (ed.), "Methods and Applications in Adaptive Control," Springer-Verlag, Berlin and New York, 1980.

34. *Automatica 20*, No. 5, Special Issue on Adaptive Control, September 1984.

35. K. J. ASTORM, P. HAGANDER, and J. STERNBY, "Zeros of Sampled Systems," *Automatica 20*, No. 1, 31-39 (1984).

36. C. E. de SOUZA, M. GEVERS, and G. C. GOODWIN, "Riccati Equations in Optimal Filtering of Nonstabilizable Systems Having Singular State Transition Matrices," *IEEE Trans. Autom. Control AC-31*, No. 9, September 1986.

37. G. C. GOODWIN, D. J. HILL, D. Q. MAYNE, and R. H. MIDDLETON, "Robust Adaptive Control: Convergence, Stability and Performance," *Tech. Report EE8544*, Electrical and Computer Engineering, University of Newcastle, Australia, 1985.

38. R. H. MIDDLETON, "Modern Continuous and Discrete Control," Ph.D. Thesis, University of Newcastle, Australia, 1986.

39. G. KREISSELMEIER, "A Robust Indirect Adaptive Control Approach," *Int. J. Control 43*, No. 1, 161-175 (1986).

Adaptive Control: A Simplified Approach

IZHAK BAR-KANA

Rafael, Haifa 31021, Israel

I. INTRODUCTION

The two main approaches in adaptive control can be classified as self-tuning regulators [1-4] and model reference adaptive control [5-8]. The impressive development of adaptive control methods over the past decade is characterized by extensions of adaptive methods to multivariable systems, by better definitions and proofs of stability and convergence properties of adaptive schemes, as well as by some extensions to nonminimum phase systems ([9-18]; see also [19] and the references therein and this volume of *Control and Dynamic Systems*).

Although adaptive control has been a dream for control systems theorists and practitioners for at least a quarter of a century [20], its implementations to realistic, difficult problems have nevertheless been very slow.

One reason for this situation may be the prior knowledge needed in order to guarantee stability of adaptive controllers. The complexity of implementation of adaptive controllers in realistic, large systems may also discourage the practitioner. Most adaptive control procedures assume that an upper bound on the order of the controlled plant and also the exact pole excess are known. This prior knowledge is needed in order to prove stability and also to implement reference models, identifiers, or observer-based controllers of about the same order as the controlled plant. Since the order of the plants in the real world may be very large or unknown, implementation of these procedures in realistic complex plants may be difficult and sometimes impossible [19,21].

Furthermore, the stability properties of such systems, when the order of the plant is underestimated, are debatable [21]. The stability problem may even be more difficult in the presence of input and output disturbances.

In an attempt to develop linear model followers that do not require observers in the feedback loop, Broussard [22] introduced a command generator tracker (CGT) approach that guarantees asymptotic tracking for step inputs with systems that can be stabilized via constant output feedback.

Using this CGT theory and Lyapunov stability-based design methods, Sobel, Kaufman, and Mabious [23,24] proposed a simplified direct model reference adaptive control algorithm for multivariable control systems. For constant input commands it was shown that this algorithm can guarantee stability and asymptotically perfect tracking if the controlled plant is "almost strictly positive real."

A system is said to be almost strictly positive real (ASPR) if there exists a constant output feedback gain matrix such that the resulting closed-loop transfer function is strictly positive real [25].

Bar-Kana [26] and Bar-Kana and Kaufman [27] extended the applicability of these simple procedures to time-variable input commands and to any system that can be stabilized via static output feedback. In these cases, the adaptive scheme can use very low-order models and does not use any observers or identifiers.

Due to their extreme simplicity compared with any other adaptive control algorithm, these procedures seem to be very attractive for applications with large-scale systems [28,29,30,31] and may thus result in a simple adaptive solution for a rather difficult problem.

However, the algorithm uses proportional *and* integral gains in order to achieve perfect tracking, and the second term may diverge whenever perfect tracking is not possible due to internal structure or due to output disturbances [21].

Therefore, in this article we present a modified adaptive algorithm that uses and completes an idea of Ioannou and Kokotovic [32] and maintains the simplicity of implementation of the basic algorithm.

It is shown that in our configuration, boundedness of all values involved in the adaptation process is guaranteed in the presence of any bounded input or output disturbances. Furthermore, the output tracking error can be controlled and arbitrarily reduced.

It is also shown that the ASPR conditions can be quite easily satisfied using parallel control in plants that can be stabilized via static or dynamic feedback.

The first part of this article presents a simplified adaptive control algorithm that can be applied if an approximate reduced-order model of the plant is known insofar as some prior raw, linear control design is available and if some stabilizable output feedback configuration is consequently known.

Although these techniques are extended, in the second part of this article, to systems in which this feedback configuration is unknown, they have their own importance due to their extreme simplicity of implementation compared with any other adaptive control schemes. This simplicity makes them especially attractive for implementation in realistic large-scale systems. We mention that a brief examination of *practical* examples that appear in the control system literature shows that most systems can be stabilized via constant output feedback. For the sake of clarity we also mention that the fictitious stabilized linear system is not required to be good. Thus, output stabilizability is only a sufficient condition that allows the *nonlinear* adaptive controller to impose upon the plant the desired behavior of the reference model.

Section II presents the basic ideas related to the positive realness and robustness of adaptive control. A scalar example shown in Section II,A is used for a intuitive introduction of the algorithms. Section II,B then defines the ASPR systems and shows how parallel feedforward can be used to satisfy the ASPR conditions if some output-stabilizing configuration is known. Section II,C presents the SISO case, for a better understanding of basic concepts. Section III presents the robust adaptive algorithms for ASPR systems. The underlying linear following control problem is analyzed in Section III,A, and the robust adaptive algorithm is the subject of Section III,B. The stability analysis of the adaptive scheme is the subject of Section III,C, and some examples are presented in Section III,D.

The generalization of the adaptive algorithm for systems in which no relevant prior knowledge is given is the subject of Section IV.

In Section IV,A a special feedforward configuration is selected in order to allow adaptive satisfaction of the positivity conditions. The linear model following formulation is given in Section IV,B. The corresponding adaptive control algorithm is presented in Section IV,C, and the stability analysis is the subject of Section IV,D. The power of these algorithms is illustrated by some difficult cases of multivariable, unstable, and nonminimum phase systems in rapidly changing environments.

II. FORMULATION OF SOME BASIC IDEAS

A. A SIMPLE ILLUSTRATIVE EXAMPLE

The following presentation is concerned with simple and robust model-reference adaptive control algorithms and with necessary conditions for implementation in realistic complex and multivariable systems.

Yet the following scalar regulator may be a useful example for a more intuitive introduction of subsequent concepts. Let the representation of the simple system be

$$\dot{x}(t) = ax(t) + bu(t) + d(t), \tag{1}$$

$$y(t) = cx(t), \tag{2}$$

where a is a positive coefficient (to make the controlled system unstable) and where d(t) is some bounded disturbance.

If the parameters are constant and known, it is not difficult to design a simple controller of the form

$$u = -ky(t) + v(t) \tag{3}$$

such that the resulting closed-loop system

$$\dot{x}(t) = (a - bkc)x(t) + bv(t) + d(t), \tag{4}$$

$$y(t) = cx(t) \tag{5}$$

is stable and also behaves in some desired way. However, if the parameters are unknown or time variable, the problem is not so simple any more. A partial solution may be proposed by using nonlinear control of the form

$$u(t) = -ky^3(t) + v(t). \tag{6}$$

In this case it is easy to see that the resulting system

$$\dot{x}(t) = [a - bkx^2(t)c^3]x(t) + bv(t) + d(t), \tag{7}$$

$$y(t) = cx(t) \tag{8}$$

is Lagrange stable (bounded) if the positivity condition $bc > 0$ is satisfied (or, at least, if the sign of bc is known). Large k values in this case may keep the output y(t) very close to the desired value [zero, for example, if $v(t) = 0$] almost independently of the magnitude of the disturbance. Of oven more importance for the following presentation is the fact that this nonlinear gain punishes high output values strongly and immediately and thus adds a strong element of robustness to the control system.

Yet this "proportional" nonlinear feedback cannot guarantee perfect following with zero tracking errors even in idealistic situations.

A still better solution may then be the use of combined proportional and "integral" adaptive controls of the form

$$k_p(t) = k_1y^2(t); \tag{9}$$

$$\dot{k}_I(t) = k_2y^2(t), \quad k_I(t) = \int \dot{k}_I(t)\, dt + k_I(0); \tag{10}$$

$$u(t) = -[k_p(t) + k_I(t)]y(t) + v(t). \tag{11}$$

It can be shown [27] for the general multivariable case that the last configuration may sometimes impose upon the plant global stability and asymptotically perfect tracking and also some very good behavior if the positivity condition is satisfied and if the disturbances are zero.

This basic adaptive algorithm still has a problem whenever disturbances are present or, in general, whenever perfect tracking is not possible. In these cases it is easily seen from (10) that the integral gain $k_I(t)$ would diverge. This difficulty can be easily overcome if, similarly to Ioanou's idea [32], we replace the integration in (10) by a first-order pole of the form

$$\dot{k}_I(t) = k_2 y^2(t) - \sigma k_I(t). \tag{12}$$

It can be shown [33,34] that the combination (9), (11), and (12) results in a robust and well-behaving control algorithm.

From the presentation above it might seem that the problems of adaptive control can be too easily overcome. However, controlling a first-order system needs no theory nor adaptive control. On the other hand, the analog of the positivity conditions in large systems leads to the concept of strictly positive real (SPR) systems so much wanted in adaptive control, even theough the reader of adaptive control literature may sometimes get the impression that SPR is only another denomination for the void class of systems.

The next section will thus try to alleviate this impression and show how to satisfy the positive realness conditions required in adaptive control.

B. THE MAIN POSITIVITY DEFINITIONS AND LEMMAS

Definition 1

Let $G_a(s)$ be an $m \times m$ transfer matrix of a continuous-time linear system. Assume that there exists a constant gain matrix $\tilde{K}_e$ such that the resulting closed-loop transfer function

$G_s(s) = [I + G_a(s)\tilde{K}_e]^{-1}G_a(s)$ is strictly positive real (SPR). In this case the original transfer matrix $G_a(s)$ is said to be almost strictly positive real (ASPR) [25].

It is well known that, if the SPR transfer matrix $G_s(s)$ has n finite poles, it may also have n finite zeros if it is proper, or n - m finite zeros if it is strictly proper, and that all poles and zeros are stable. Consequently, $G_a(s)$ must also have n zeros or n - m zeros, correspondingly. The poles are not necessarily stable, however, all zeros must be placed in the left half plane (minimum phase).

The class of ASPR systems will be the most favored class for the subsequent application of simplified adaptive control; however, it may be very restrictive. Therefore we first extend its limits to systems that can be stabilized via constant output feedback with the following lemma.

Lemma 1

Let $G_p(s)$ be the transfer matrix of a continuous-time linear system. $G_p(s)$ is not necessarily stable or minimum phase. Let us assume that there exists a nonsingular constant feedback matrix K_y such that the resulting closed-loop transfer matrix $G_c(s) = [I + G_p(s)K_y]^{-1}G_p(s)$ is asymptotically stable. In this case there exists a nonsingular feedforward gain matrix, for example, $D = K^{-1}$, such that the new open-loop transfer matrix $G_a(s) = D + G_p(s)$ is ASPR [25].

The applicability of these concepts and of the adaptive control techniques based on them can be further extended to include systems that need dynamic feedback in order to reach stabilizability. However, as we will show later, the ideas presented above are sufficient for implementation of a simple and robust adaptive control algorithm with many practical control systems.

We note first that direct (constant) feedforward in combination with direct feedback may create difficulties in implementation of adaptive control schemes due to the algebraic input-output-input loop that appears. This difficulty can be eliminated since $G_a(s)$ remains ASPR if the constant feedforward D is replaced by a first-order dynamics of the form

$$D(s) = D/(1 + s/s_0) \tag{13}$$

and if s_0 is (selected) sufficiently large. The augmented open-loop transfer matrix of the plant is thus

$$G_a(s) = D/(1 + s/s_0) + G_p(s). \tag{14}$$

For the subsequent presentation we assume that D and s_0 are selected a priori and that the adaptive algorithm is required to control the augmented ASPR plant (14). The most preferred plants for these applications are those systems that become stable or maintain stability with high feedback gains K_y. In this case according to Lemma 1, the feedforward gain D may be very small, such that practically, $G_a(s) \simeq G_p(s)$.

The stabilizability via constant output feedback was assumed above both because of its simplicity and because of its surprisingly wide applicability. This last point can be checked by a brief examination of the examples appearing in the control system literature.

Yet, for the sake of generality, we extend the applicability of the ASPR concept through the following lemma.

Lemma 2

Let us assume that $G_p(s)$ is an $m \times m$ transfer function, not necessarily stable or minimum phase. Let H(s) be a dynamic $m \times m$ output feedback of some order $p < n$ such that the resulting closed-loop transfer matrix $G_1(s) = [I + G_p(s)H(s)]^{-1}G_p(s)$ is asymptotically stable. In this case the augmented open-loop transfer matrix

$$G_a(s) = G_p(s) + H^{-1}(s) \tag{15}$$

is ASPR if $H^{-1}(s)$ is ASPR [25].

C. THE SISO CASE

The multivariable case presented in the previous section may confuse the reader. Therefore, in this section we present the SISO case in a form that may help formation of a more intuitive representation of the ASPR concept.

A strictly proper ASPR transfer function, in the SISO case, is minimum phase, the degree of its denominator polynomial (not necessarily Hurwitz) is n, and the degree of its Hurwitz numerator polynomial is n - 1.

Let $G_p(s) = B_p(s)/A_p(s)$ be a SISO transfer function, not necessarily stable or minimum phase, and having any pole excess. Let $H(s) = Q(s)/P(s)$ be some output-stabilizing feedback such that the closed-loop transfer matrix $G_c(s) = B_p(s)P(s)/[A_p(s)P(s) + B_p(s)Q(s)]$ is asymptotically stable. Then all roots of the denominators $D(s) = A_p(s)P(s) + B_p(s)Q(s)$ are placed in the left half plane.

Let us consider now the augmented open-loop transfer function $G_a(s) = G_p(s) + H^{-1}(s) = [A_p(s)P(s) + B_p(s)Q(s)]/A_p(s)Q(s)$.

It is clear that all zeros of $G_a(s)$ are placed in the left half plane and that $G_a(s)$ is therefore minimum phase. Furthermore, if the pole excess of $H^{-1}(s)$ is selected to be 1, it is easy to see that the pole excess of $G_a(s)$ will also be 1, and thus, that $G_a(s)$ is ASPR.

The application of the lemmas is immediate if some approximate low-order representation of the plant is available and if a raw linear control design exists insofar as some stabilizing output feedback can be given. It is worth mentioning that the fictitious closed-loop system is only required to be stable, not "good." Thus, the output stabilizability of the controlled plant is only a sufficient condition that allows the subsequent adaptive control algorithms to impose upon the augmented controlled plant the desired behavior of the model. However, it is useful to maintain the gains of $H^{-1}(s)$ as small as possible, such that $G_a(s) = G_p(s) + H^{-1}(s) \simeq G_p(s)$. Otherwise, special care must be taken in order to eliminate the bias introduced by the supplementary feedforward [26,35]. Let the augmented ASPR-controlled plant (15) have the following state-space representation (Fig. 1):

$$\dot{x}(t) = Ax(t) + Bu(t), \tag{16}$$

$$y(t) = Cx(t). \tag{17}$$

It is useful, for the following analysis, to also present the fictitious SPR system that could be obtained through fictitious constant feedback $\tilde{K}_e$:

$$\dot{x}(t) = A_c x(t) + Bu(t), \tag{18}$$

$$y(t) = Cx(t), \tag{19}$$

where

$$A_c = A - B\tilde{K}_e C. \tag{20}$$

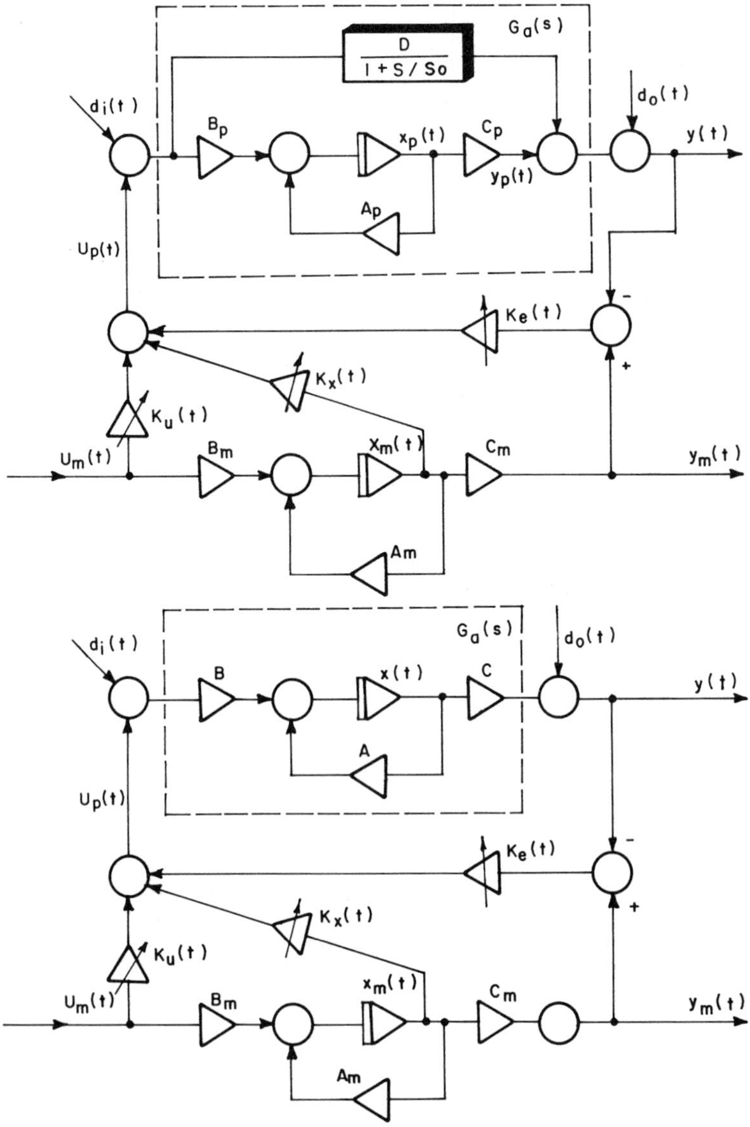

Fig. 1. Representation of the two equivalent adaptive control configurations.

III. SIMPLIFIED ADAPTIVE CONTROL IN ASPR SYSTEMS

A. *FORMULATION OF THE LINEAR LOW-ORDER MODEL FOLLOWING PROBLEM*

The output of the large, augmented ASPR plant (16)-(17) is required to track the output of a low-order model reference:

$$\dot{x}_m(t) = A_m x_m(t) + B_m u_m(t), \tag{21}$$

$$y_m(t) = C_m x_m(t). \tag{22}$$

The model incorporates within it the desired input-output behavior of the plant, but is free otherwise. It is especially permitted that

$$\dim[x] \gg \dim[x_m]. \tag{23}$$

For example, the output of a system of order (say) 100 may be required to track the output of a well-designed first- or second-order model. Furthermore, the order of the plant is unknown and not available for implementation. These difficult assumptions are introduced from the start because in the author's opinion they are representative for all practical control design problems.

To this end we want to use simple controllers of the form [23]

$$u_p(t) = \tilde{K}_{e_y}[y_m(t) - y(t)] + \tilde{K}_{x_m} x_m(t) + \tilde{K}_{u_m} u_m(t) \tag{24}$$

such that the plant output ultimately tracks the output of the model or

$$y(t) = y_m(t). \tag{25}$$

In this case the control $u_p(t)$ becomes the "ideal" control $u_p^*(t)$ of the form

$$u_p^*(t) = \tilde{K}_{x_m} x_m(t) + \tilde{K}_{u_m} u_m(t). \tag{26}$$

It is important to note that for the linear model following problem, this ideal situation is not necessarily possible. In other words, it is not necessarily guaranteed that solutions of (16)-(17) can satisfy (25) when excited by the ideal input command (26). Therefore, for the subsequent adaptive control presentation, we only assume that there exists a fictitious target system of the form

$$\dot{x}^*(t) = A^* x^*(t) + B^* u^*(t), \tag{27}$$

$$y^*(t) = Cx^*(t) \tag{28}$$

that satisfies (25) when forced by the command (26). Note that although (27) is assumed to be of the same (unknown) order as (16), these two systems may be entirely different, and we only assume that their output-measuring matrices in (17) and (28) are identical.

In order to check the existence of bounded ideal target trajectories, we assume for convenience and with some irrelevant restriction of generality that $x^*(t)$ has the following form:

$$x^*(t) = Xx_m(t). \tag{29}$$

The output tracking condition then becomes

$$y^*(t) = Cx^*(t) = CXx_m(t) = C_m x_m(t) = y_m(t) \tag{30}$$

or

$$CX = C_m. \tag{31}$$

Equation (31) has a solution for the matrix X if

$$\text{rank}[C \quad C_m] = \text{rank}\ C, \tag{32}$$

which is satisfied, in general, especially under (23).

We will show subsequently that, if (31) is satisfied, then the nonlinear adaptive control algorithm can lead the plant (16)-(17) arbitrarily close to perfect tracking. [We will assume for convenience that (31) *is* satisfied. However, it is worth noting that the subsequent presentation would be valid if we replaced the exact solution of (31) by the least mean square solution.]

B. THE ROBUST ADAPTIVE CONTROL ALGORITHM IN ASPR SYSTEMS

In a realistic environment the augmented ASPR-controlled plant, $G_a(s)$ (Fig. 1), has the following representation:

$$\dot{x}(t) = Ax(t) + Bu(t) + d_i(t), \tag{33}$$

$$y(t) = Cx(t) + d_o(t), \tag{34}$$

where $d_i(t)$ and $d_o(t)$ are any bounded input or output disturbances.

We now define the state error as

$$e_x(t) = x^*(t) - x(t) \tag{35}$$

and the output error as

$$e_y(t) = y_m(t) - y(t) = y^*(t) - y(t) = Ce_x(t) - d_o(t). \tag{36}$$

Since the controlled plant is unknown, we replace the constant gains in (24) by time-variable adaptive gains as follows:

$$u_p(t) = K_{e_y}(t)e_y(t) + K_{x_m}(t)x_m(t) + K_{u_m}(t)u_m(t)$$

$$= K(t)r(t), \tag{37}$$

where

$$K(t) = [K_{e_y}(t) \quad K_{x_m}(t) \quad K_{u_m}(t)], \tag{38}$$

$$r^T(t) = \left[e_y^T(t) \quad x_m^T(t) \quad u_m^T(t)\right]. \tag{39}$$

The output error $e_y(t)$ is then used to calculate the adaptive gains that are obtained as a combination of proportional and integral terms of the form

$$K(t) = K_p(t) + K_I(t), \tag{40}$$

$$K_p(t) = e_y(t) r^T(t) \overline{T}, \tag{41}$$

$$\dot{K}_I(t) = \left[e_y(t) r^T(t) - \sigma K_I(t)\right] T, \tag{42}$$

where $\overline{T}$ and T are selected (constant or time-variable) positive-definite adaptation coefficient matrices. The adaptive gains (40)-(42) use and modify an idea of Ioannou and Kokotovic [32] and the σ term in (42) is thus introduced in order to guarantee robustness in the presence of parasitic disturbances. It will be shown that our configuration (40)-(42) guarantees global stability with any bounded input *and* output disturbance. We will also show that the proportional term (41) facilitates direct control of the output error $\dot{e}_y(t)$. This way, the output error can be ultimately reduced almost arbitrarily. (The treatment of the perfect tracking case in idealistic conditions is given in [27].)

At this stage it is hoped that the reader is aware of the extreme simplicity and of the low order of the algorithm (37)-(42) compared with any other adaptive control algorithm, because this may be the main contribution of our approach. It is definitely not the general solution for the general problem;

however, it may provide a simple and well-performing solution for some realistic complex problems.

By differentiating $e_x(t)$ in (19), we obtain

$$\dot{e}_x(t) = \dot{x}^* - \dot{x}(t) = \dot{x}^*(t) - Ax^*(t) + Ax^*(t) - \dot{x}(t). \tag{43}$$

After substituting the appropriate expressions in (43) and after some algebraic manipulations, we obtain the following differential equation of the state error:

$$\dot{e}_x(t) = A_c e_x(t) - B(K(t) - \tilde{K})r(t) - F(t), \tag{44}$$

where A_c was defined in (20) and where

$$\tilde{K} = \left[\tilde{K}_{e_y} \quad \tilde{K}_{x_m} \quad \tilde{K}_{u_m}\right], \tag{45}$$

$$F(t) = \left(AX - XA_m + B\tilde{K}_x\right)x_m(t) - \left(XB_m - B\tilde{K}_u\right)u_m(t)$$
$$+ B\tilde{K}_e d_o(t) + d_i(t). \tag{46}$$

C. STABILITY ANALYSIS

In this section we prove the following theorem of stability for the proposed adaptive control procedure.

Theorem 1

Let us assume that the (eventually) augmented, controlled plant (16)-(17) is ASPR, that is, that there exists a constant feedback gain matrix $\tilde{K}_e$, unknown and not needed for implementation, such that the resulting fictitious closed-loop system (18)-(19) is SPR.

Then all values involved in the control of the augmented process (33)-(34) via the adaptive procedure (37)-(42), namely, states, gains, and errors, are bounded. Furthermore, the output tracking error $e_y(t)$ can be directly controlled and thus ultimately reduced via the adaptation coefficient $\bar{T}$.

Proof. If the undisturbed plant (16)-(17) is ASPR, the fictitious closed-loop SPR configuration (18)-(19) satisfies the following relations [36]:

$$PA_c + A_c^T P = -Q < 0, \tag{47}$$

$$PB = C^T, \tag{48}$$

where P and Q are some unknown positive-definite matrices.

The following quadratic Lyapunov function is used to prove stability of the adaptive system described by (42) and (44):

$$V(t) = e_x^T(t) P e_x(t) + \mathrm{tr}\left[\left(K_I(t) - \tilde{K}\right) T^{-1} \left(K_I(t) - \tilde{K}\right)^T\right], \tag{49}$$

where P is the positive-definite matrix defined in (47)-(48) and T the adaptation coefficient defined in (42).

If relations (47)-(48) are satisfied, the derivative of the Lyapunov function (49) becomes (Appendix A)

$$\begin{aligned} \dot{V}(t) = & -e_x^T(t) Q e_x(t) - 2\sigma\, \mathrm{tr}\left[\left(K_I(t) - \tilde{K}\right)\left(K_I(t) - \tilde{K}\right)^T\right] \\ & - 2e_y^T(t) e_y(t) r^T(t) \bar{T} r(t) - 2\sigma\, \mathrm{tr}\left[\left(K_I(t) - \tilde{K}\right)\tilde{K}^T\right] \\ & - 2e_x^T(t) P F(t) - 2d_o^T(t)(K(t) - \tilde{K}) r(t). \end{aligned} \tag{50}$$

We can see that if either $e_x(t)$, $K_I(t)$, or $e_y(t)$ becomes large, one or more of the first three terms in (50) become negative and dominant, such that $\dot{V}(t)$ becomes negative. The quadratic form (49) or V(t) then guarantees that all states, gains, and errors are bounded. These values are ultimately bounded by the set

$$\left\{e_x, K_I \,\middle|\, \dot{V}(t) = 0\right\}. \tag{51}$$

We note that the third term in (50), which is a direct result of the proportional gain (41), is negative-definite quartic with respect to the output error and proportional to the adaptation coefficient $\bar{T}$. Thus, by appropriate selection of $\bar{T}$, the output error can be controlled and reduced almost arbitrarily, as a trade-off between large adaptation coefficients and large errors. We must mention that the effect of this term is so strong that in some simulations it is difficult to be convinced that the other positivity conditions are necessary at all.

It is also worth mentioning that the constant $\bar{T}$ can be replaced by a time-variable coefficient $\bar{T}(t)$ without affecting the stability analysis. Thus this coefficient may be properly selected in practice in order to obtain both smooth time response (with small coefficients at the start, when the errors are usually large) and small final tracking errors (with larger coefficients).

D. *EXAMPLES*

The representations of a large structure (plate) and an unstable helicopter are used to illustrate the power of the robust adaptive algorithm. No information about the plants was used for implementation of the adaptive schemes, and all adaptive gains were initially zero.

Example 1

A totally undamped 20-state (10-mode) two-input, two-output representation of a 1.5 × 1.5 m, 1.0 mm thick simply supported aluminum plate [37,26,28,29] is used here for robust control with square-wave input commands. Position plus velocity feedback is used in order to stabilize and control the shape of the plate.

The plant has the following representation:

$$\dot{x}_p(t) = \begin{bmatrix} 0 & I_{10} \\ -\Lambda_{10} & 0 \end{bmatrix} x_p(t) + \begin{bmatrix} 0 \\ B_{10} \end{bmatrix} u_p(t) + d_i(t), \tag{52}$$

$$y_p(t) = [C_{10} \quad C_{10}]x_p(t) + d_o(t), \tag{53}$$

where

$$\Lambda_{10} = \mathrm{diag}(414.428,\ 2590.179,\ 2590.179,\ 6630.85,\ 10360.716,\ 10360.716,\ 17409.982,\ 17409.982,\ 29855.553,\ 29855.553), \tag{54}$$

$$C_{10} = B_{10}^T = \begin{bmatrix} 0.326 & 0.4367 & -0.1357 & 0.1816 & -0.270 & 0.2581 \\ & & -0.107 & -0.361 & 0.248 & -0.0913 \\ 0.4246 & 0.0888 & 0.2625 & -0.0549 & -0.2625 & -0.406 \\ & & -0.250 & -0.0549 & -0.4246 & 0.1737 \end{bmatrix}. \tag{55}$$

The model reference is a damped representation of the first two modes of the plate:

$$\dot{x}_m(t) = \begin{bmatrix} 0 & I_2 \\ -\Lambda_2 & -2z\sqrt{\Lambda_2} \end{bmatrix} x_m(t) + \begin{bmatrix} 0 \\ B_2 \end{bmatrix} u_m(t), \tag{56}$$

$$y_m(t) = [C_2 \quad C_2]x_m(t), \tag{57}$$

where $z = 0.8$ and

$$\Lambda_2 = \mathrm{diag}(414.428,\ 2520.179), \tag{58}$$

$$C_2 = B_2^T = \begin{bmatrix} 0.326 & 0.4367 \\ 0.4246 & 0.0888 \end{bmatrix}. \tag{59}$$

The plant is almost (not strictly) positive real [26]. However, in the presence of disturbances the plant is required to be ASPR; therefore, we (heuristically) select the supplementary feedforward of the form

$$y_s(s) = \frac{\begin{bmatrix} 0.001 & 0 \\ 0 & 0.001 \end{bmatrix}}{1 + s/500} u_s(s). \tag{60}$$

The disturbances are

$$d_i(t) = \begin{bmatrix} 5 \\ 10\sin(6t) \end{bmatrix} \tag{61}$$

and

$$d_o(t) = \begin{bmatrix} 0.003 \\ 0.004\ \sin(10t) \end{bmatrix} \tag{62}$$

The adaptation coefficients were $T = \bar{T} = 10{,}000 I_8$.

The simulation represented in Figs. 2a-d was started with $\sigma = 0.0$ in clear environment and shows good output following. Note Figs. 2a-b show the actual plant outputs $y_p(t)$ [not the augmented output $y(t)$, which is only of interest for stability]. It can be seen that the gains $K_{e_{11}}(t)$ and $K_{e_{22}}(t)$ show the tendency to increase whenever the output tracking is not perfect.

At $t = 1$ sec the value of σ was changed to $\sigma = 0.001$ and the effect on the gain values is evident. Results of simulation show good output following with (small) bounded errors. Input *and* output disturbances were then introduced at $t = 3.6$ sec, and we obtained similar results, although in this case the output errors and the gains were larger.

Example 2

The second example is an oscillatory representation of an unstable helicopter [38], with the only difference that our example is also nonminimum phase in order to make it even more difficult. The transfer function of the plant is

$$y_p(s) = \frac{s - 0.03}{(1 + 2s)(s^2 - 0.35s + 0.15)}\, u_p(s). \tag{63}$$

The plant is required to track the output of the *first-order* model

$$y_m(s) = \frac{1}{1 + s/3}\, u_m(s). \tag{64}$$

The supplementary feedforward is selected as

$$D(s) = \frac{0.001}{1 + s/100}, \tag{65}$$

and the adaptation coefficients are $T = \bar{T} = 10{,}000 I_3$.

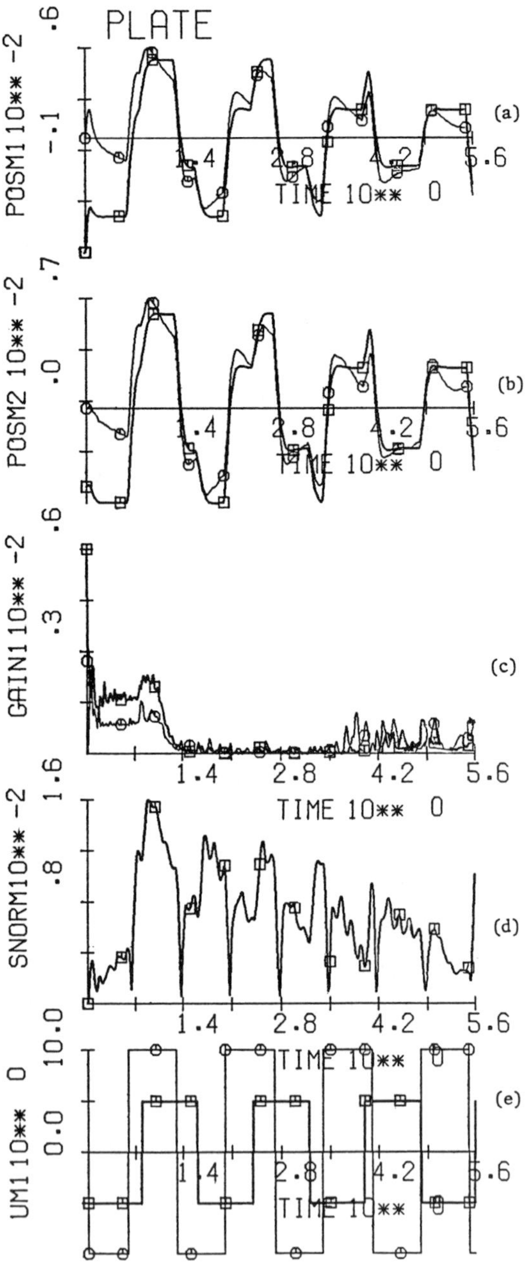

Fig. 2. Example 1: a, output $y_{m_1}(t)$, $y_{p_1}(t)$; b, output $y_{m_2}(t)$, $y_{p_2}(t)$; c, gains $K_{e_{11}}(t)$, $K_{e_{22}}(t)$; d, state vector norm; e, input $u_{m_1}(t)$, $u_{m_2}(t)$.

The results of simulation are shown in Figs. 3a-d, with $\sigma = 0.0$ for 15 sec in clean environment, and show good output tracking; however, the gains increase steadily due to the tracking errors. The same case is continued with $\sigma = 0.001$ at $t = 15$ sec and shows quite good behavior and immediate reduction of the gain.

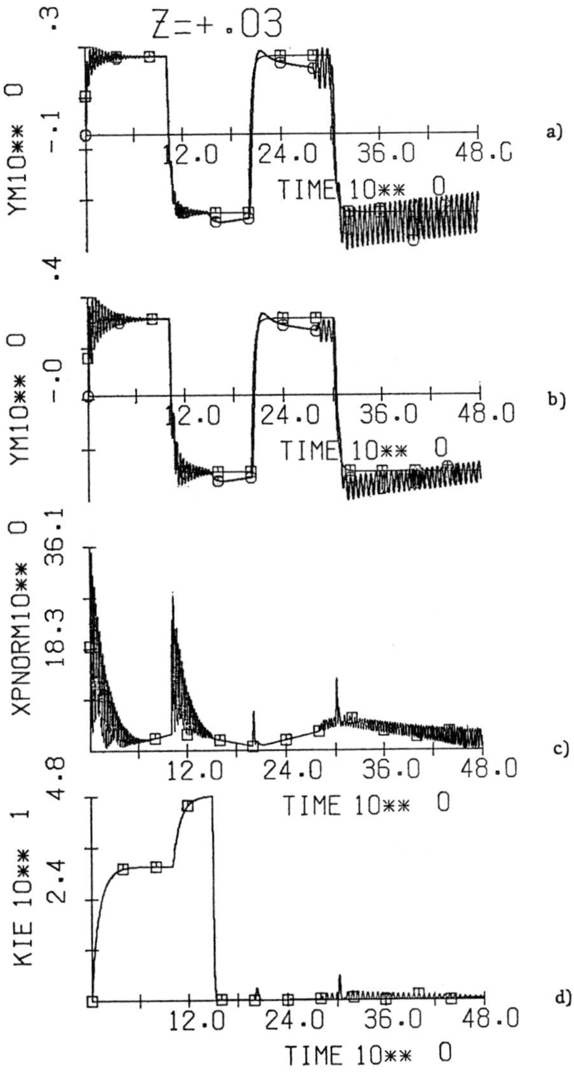

Fig. 3. Example 2: a, output $y_m(t)$, augmented output $y(t)$; b, output $y_m(t)$, $y_p(t)$; c, state vector norm; d, gain $K_{Ie}(t)$.

A rather large output disturbance was then added at t = 28 sec,

$$d_o = 0.08 \sin 10t, \tag{66}$$

and results of this simulation show boundedness of all values involved in the adaptation process.

IV. GENERALIZATION OF THE SIMPLIFIED ADAPTIVE ALGORITHM

If the prior knowledge about the controlled plant is very poor, one cannot assume that the stabilizing feedback H(s) and the consequent feedforward dynamics $H^{-1}(s)$ can be given a priori. Therefore, a preselected fixed feedforward configuration may not always guarantee ASPR properties and, on the other hand, may add a too large parallel bias when not necessarily needed.

The next step, which is presented in this section, is an attempt to calculate the parallel feedforward adaptively. It is assumed only that the controlled plant is output stabilizable via some output feedback dynamics of order not larger than some known number p.

A. THE POSITIVE REAL CONFIGURATION WHEN ALL COEFFICIENTS ARE KNOWN

Following [25] and Lemma 2, we present a procedure that can be used to achieve a strictly positive real configuration.

A state-space representation of the controlled plant $G_p(s)$ is

$$\dot{x}_p(t) = A_p x_p(t) + B_p u_p(t), \tag{67}$$

$$y_p(t) = C_p x_p(t), \tag{68}$$

and a basic supplementary feedforward $H^{-1}(s)$ is selected, for convenience of the subsequent adaptive procedure, as

$$\dot{x}_s(t) = A_s x_s(t) + B_s u_s(t), \tag{69}$$

$$y_s(t) = C_s x_s(t), \tag{70}$$

where

$$A_s = \begin{bmatrix} 0 & I_m & 0 & \cdots & 0 \\ 0 & 0 & I_m & \cdots & 0 \\ \hline 0 & 0 & 0 & \cdots & I_m \\ 0 & 0 & 0 & \cdots & 0 \end{bmatrix}, \quad B_s = \begin{bmatrix} 0 \\ 0 \\ \vdots \\ 0 \\ I_m \end{bmatrix} \tag{71}$$

and where C_s is selected a priori to establish the desired zero position of the feedforward (69)-(70). The output measuring equation is

$$y(t) = y_p(t) + y_s(t) = C_p x_p(t) + C_s x_s(t). \tag{72}$$

The control is defined by

$$u_p(t) = -\tilde{K}_{e_y} y(t) + u(t), \qquad u_s(t) = \tilde{K}_{x_s} x_s(t) + \tilde{K}_{u_p} u_p(t). \tag{73}$$

The closed-loop system is then

$$\begin{bmatrix} \dot{x}_p(t) \\ \dot{x}_s(t) \end{bmatrix} = A \begin{bmatrix} x_p(t) \\ x_s(t) \end{bmatrix} + Bu(t), \qquad y(t) = C \begin{bmatrix} x_p(t) \\ x_s(t) \end{bmatrix}, \tag{74}$$

where

$$A = \begin{bmatrix} A_p - B_p\tilde{K}_{e_y}C_p & -B_p\tilde{K}_{e_y}C_s \\ -B_s\tilde{K}_{u_p}\tilde{K}_{e_y}C_p & A_s - B_s\tilde{K}_{x_s} - B_s\tilde{K}_{u_p}\tilde{K}_{e_y}C_s \end{bmatrix}, \tag{75}$$

$$B = \begin{bmatrix} B_p \\ B_s\tilde{K}_{u_p} \end{bmatrix}, \qquad C = [C_p \quad C_s].$$

By definition, the fictitious system (74) is strictly positive real; therefore, there exist two positive-definite matrices P and Q such that the following relations are satisfied [36]:

$$PA + A^TP = -Q < 0, \tag{76}$$

$$PB = C^T. \tag{77}$$

Relations (76)-(77) will be necessary for the following stability analysis.

B. THE LINEAR MODEL FOLLOWING FORMULATION

The output of the basic augmented system (67)-(72) is required to track the output of the low-order model (21)-(22).

The controls have the following configuration:

$$u_p(t) = \tilde{K}_{e_y}[y_m(t) - y(t)] + \tilde{K}_{x_m}x_m(t) + \tilde{K}_{u_m}u_m(t) = \tilde{K}r(t), \tag{78}$$

$$u_s(t) = \tilde{K}_{x_s}x_s(t) + \tilde{K}_{u_p}u_p(t), \tag{79}$$

where

$$\tilde{K} = \begin{bmatrix}\tilde{K}_{e_y} & \tilde{K}_{x_m} & \tilde{K}_{u_m}\end{bmatrix}, \tag{80}$$

$$r^T(t) = \begin{bmatrix}(y_m(t) - y(t))^T & x_m^T(t) & u_m^T(t)\end{bmatrix}. \tag{81}$$

Since the model may be of very low order compared with the controlled plant, we first verify if perfect output tracking is possible. If the augmented plant is perfectly tracking, then $y(t) = y_m(t)$, the control (78)-(79) becomes the ideal control

$$u_p^*(t) = \tilde{K}_{x_m}x_m(t) + \tilde{K}_{u_m}u_m(t), \quad u_s^*(t) = \tilde{K}_{x_s}x_s^*(t) + \tilde{K}_{u_p}u_p^*(t), \tag{82}$$

and the plant moves along some bounded ideal trajectories that make perfect tracking possible.

In order to verify existence of bounded target trajectories, we assume for convenience, and with some irrelevant restriction of generality, that they have the following form:

$$x_p^*(t) = X_p x_m(t), \quad x_s^*(t) = X_s x_m(t) \tag{83}$$

such that the output tracking condition

$$y_p^*(t) = C_p x_p^*(t) = C_m x_m(t) = y_m(t), \quad y_s^*(t) = C_s x_s^*(t) = 0 \tag{84}$$

is satisfied.

Substituting (19) into (20) gives

$$C_p X_p = C_m, \quad C_s X_s = 0. \tag{85}$$

Equations (85) have many more variables than equations and have (at least) a solution for X_p and X_s, in general. For this reason the restriction of generality (83) was called "irrelevant."

Note that $x_p^*(t)$ and $x_s^*(t)$ are "target" trajectories and are not necessarily solutions of (67) and (69). The subsequently proposed adaptive algorithm will attempt to drive the plant arbitrarily close to these target trajectories.

C. THE ADAPTIVE CONTROL ALGORITHM

In a realistic environment the augmented controlled plant is represented by

$$\dot{x}_p(t) = A_p x_p(t) + B_p u_p(t) + E_p d_i(t), \tag{86}$$

$$\dot{x}_s(t) = A_s x_s(t) + B_s u_p(t) + E_s d_i(t), \tag{87}$$

and the output measurement equation is

$$y(t) = y_p(t) + y_s(t) = C_p x_p(t) + C_s x_s(t) + d_o(t), \tag{88}$$

where A_s, B_s, and C_s were defined in (69)-(71) and where $d_i(t)$ and $d_o(t)$ are any bounded input and output disturbances.

We now define the state errors as

$$e_p(t) = x_p^*(t) - x_p(t), \quad e_s(t) = x_s^*(t) - x_s(t), \tag{89}$$

and the output error is then

$$e_y(t) = y_m(t) - y(t) = y^*(t) - y(t) = [C_p \quad C_s]\begin{bmatrix} e_p \\ e_s \end{bmatrix} - d_o(t). \tag{90}$$

Since the controlled plant is unknown, we replace the constant gains in (78)-(79) by time-variable adaptive gains as follows:

$$u_p(t) = K_{e_y}(t)e_y(t) + K_{x_m}(t)x_m(t) + K_{u_m}(t)u_m(t) = K(t)r(t), \tag{91}$$

$$u_s(t) = K_{x_s}(t)x_s(t) + K_{u_p}u_p(t) = K_s(t)r_s(t), \tag{92}$$

where again

$$r^T(t) = \left[e_y^T(t) \quad x_m^T(t) \quad x_u^T(t)\right], \tag{93}$$

$$r_s^T(t) = [x_s(t) \quad u_p(t)], \tag{94}$$

$$K(t) = [K_{e_y}(t) \quad K_{x_m}(t) \quad K_{u_m}(t)], \tag{95}$$

$$K_s(t) = [K_{x_s}(t) \quad K_{u_p}(t)]. \tag{96}$$

Now we use both output values, $e_y(t)$ *and* $y_s(t)$, that we want to minimize, and the adaptive gains are obtained as a combination of proportional and integral gains of the form (Fig. 4)

$$K(t) = K_p(t) + K_I(t), \tag{97}$$

$$K_p(t) = e_y(t)r^T(t)\bar{T}, \tag{98}$$

$$\dot{K}_I(t) = \left[e_y(t)r^T(t) - \sigma K_I(t)\right]T, \tag{99}$$

and also

$$K_s(t) = K_{sp}(t) + K_{sI}(t), \tag{100}$$

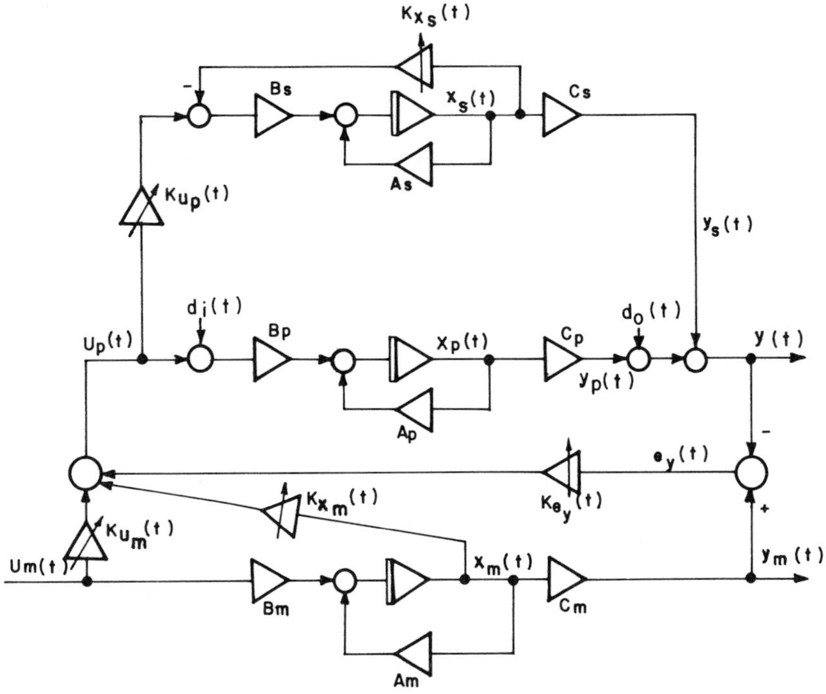

Fig. 4. The complete adaptive control system representation.

$$K_{sp}(t) = e_{y_s}(t) r_s^T(t) \overline{T}_s, \tag{101}$$

$$\dot{K}_{sI}(t) = \left[e_{y_s}(t) r_s^T(t) - \sigma K_{sI}(t)\right] T_s, \tag{102}$$

where T, $\overline{T}$, T_s, and $\overline{T}_s$ are selected (constant or time-variable) positive-definite adaptation coefficient matrices and where we defined for convenience

$$e_{y_s}(t) = y_s^*(t) - y_s(t) = -y_s(t). \tag{103}$$

The σ terms in (99) and (102) are again introduced in order to guarantee robustness in the presence of parasitic disturbances.

It is worth mentioning that (100)-(102) is not the most general procedure for the feedforward dynamics, since it only adjusts the adaptive gains $K_s(t)$ to the a priori selected matrix

C_s. In other words, the adaptive algorithm calculates the gains and the poles of the feedforward, while its zeros are preselected. It may be better to also calculate C_s adaptively, but this needs more research.

If we define

$$e_x(t) = \begin{bmatrix} e_p(t) \\ e_s(t) \end{bmatrix}, \tag{104}$$

then the differential equation of the state error is

$$\dot{e}_x(t) = \begin{bmatrix} \dot{e}_p(t) \\ \\ \dot{e}_s(t) \end{bmatrix} = \begin{bmatrix} \dot{x}_p^*(t) - \dot{x}_p(t) \\ \\ x_s^*(t) - x_s(t) \end{bmatrix}$$

$$= \begin{bmatrix} \dot{x}_p^*(t) - A_p\dot{x}_p^*(t) + A_p x_p^*(t) - \dot{x}_p(t) \\ \\ \dot{x}_s^*(t) - A_s x_s^*(t) + A_s x_s^*(t) - \dot{x}_s(t) \end{bmatrix}. \tag{105}$$

After substituting the appropriate expressions in (105) and after some algebraic manipulations, we obtain

$$\dot{e}_x(t) = Ae_x(t) - B[K(t) - \tilde{K}]r(t) + \begin{bmatrix} 0 \\ B_s \end{bmatrix}$$

$$\times \left[K_s(t) - \tilde{K}_s\right]x_s(t) - F(t), \tag{106}$$

where A and B were defined in (75) and where

$$F(t) = \begin{bmatrix} \left(A_pX_p - X_pA_m + B_p\tilde{K}_{x_m}\right)x_m(t) - \left(X_pB_m - B_p\tilde{K}_{u_m}\right)u_m(t) \\ \qquad + E_pd_i(t) + B_p\tilde{K}_ed_o(t) \\ \left(A_sX_s - X_sA_m + B_s\tilde{K}_{u_p}\tilde{K}_x\right)x_m(t) - \left(X_sB_m - B_s\tilde{K}_{u_p}\tilde{K}_{u_m}\right)u_m(t) \\ \qquad + E_sd_i(t) + B_s\tilde{K}_{u_p}\tilde{K}_ed_o(t) \end{bmatrix}. \tag{107}$$

The output equation (90) then becomes

$$e_y(t) = Ce_x(t) - d_o(t). \tag{108}$$

D. STABILITY ANALYSIS

The following quadratic Lyapunov function is used to analyze the stability of the adaptive system described by (99), (102), and (106):

$$V(t) = e_x^T(t)Pe_x(t) + \mathrm{tr}\left[\left(K_I(t) - \tilde{K}\right)T^{-1}\left(K_I(t) - \tilde{K}\right)^T\right] + \mathrm{tr}\left[\left(K_{sI}(t) - \tilde{K}_s\right)T_s^{-1}\left(K_{sI}(t) - \tilde{K}_s\right)^T\right], \tag{109}$$

where P is the positive-definite matrix defined in (76)-(77) and T and T_s are the adaptation coefficient matrices defined in (99) and (102).

If the relations (76)-(77) are satisfied, the derivative of the Lyapunov function (109) becomes (Appendix B)

$$\begin{aligned}\dot{V}(t) = &-e_x^T(t)Qe_x(t) - 2\sigma\ \mathrm{tr}\left[\left(K_I(t) - \tilde{K}\right)\left(K_I(t) - \tilde{K}\right)^T\right]\\ &- 2\sigma\ \mathrm{tr}\left[\left(K_{sI}(t) - \tilde{K}_s\right)\left(K_{sI}(t) - \tilde{K}_s\right)^T\right]\\ &- 2e_y^T(t)e_y(t)r^T(t)\overline{T}r(t) - 2y_s^T(t)y_s(t)r_s^T(t)\overline{T}_s r_s(t)\\ &- 2\sigma\ \mathrm{tr}\left[\left(K_I(t) - \tilde{K}\right)\tilde{K}^T\right] - 2\sigma\ \mathrm{tr}\left[\left(K_{sI}(t) - \tilde{K}_s\right)\tilde{K}_s^T\right]\\ &- 2e_x^T(t)PF(t) - 2d_o^T(t)(K(t) - \tilde{K})r(t)\\ &+ 2e_p^T(t)P_{12}B_s\left[K_s(t) - \tilde{K}_s\right]r_s(t)\\ &+ 2e_s^T(t)\left(P_{22}B_s - C_s^T\right)\left[K_s(t) - \tilde{K}_s\right]r_s(t),\end{aligned} \tag{110}$$

where P_{12} and P_{22} are defined in Appendix B.

$\dot{V}(t)$ consists of five negative-definite or semidefinite terms and other nondefinite terms. The stability properties of the adaptive control system depend on the relative magnitude of these terms. It can be seen that if either $e_x(t)$, $K_I(t)$, $K_{sI}(t)$, $e_y(t)$, or $y_s(t)$ becomes large, the first five negative terms become dominant over all but the last two terms. This is possible due to the contribution of the fourth and fifth terms in (110), which are negative-definite quartic with respect to the output errors $e_y(t)$ and $y_s(t)$.

Still, there remains the problem of the last two terms. First, the last term may very well vanish, because the special selection of the feedforward allows existence of some matrix P_{22} such that $P_{22}B_s = C_s^T$.

However, in general we have no reason to believe that $P_{12}B_s = 0$. Therefore, the stability properties depend on the following condition: if the next to the last term, namely, $2e_p^T P_{12} B_s \left[K_s(t) - \tilde{K}_s\right] r_s(t)$, is also dominated by the first negative-definite terms, then for large enough magnitudes of the controlled values, $\dot{V}(t)$ becomes negative. The quadratic form of the Lyapunov function $V(t)$ then guarantees that all these values are bounded and ultimately enclosed within the set $\left\{e_x(t),\ K_I(t),\ K_{sI}(t) \,|\, \dot{V}(t) \equiv 0\right\}$. Note the special role of the fourth and fifth terms, namely, $-2e_y^T(t)e_y(t)r^T(t)\overline{T}r(t)$ and $-2y_s^T(t)y_s(t)r_s^T(t)\overline{T}_s r_s(t)$: these negative definite terms with respect to the output errors $e_y(t)$ and $y_s(t)$ allow direct control of the output error, through appropriate selection of the adaptation coefficients $\overline{T}$ and $\overline{T}_s$. It is believed that these terms also help in achieving negativeness of $\dot{V}(t)$ whenever the adaptive system tends to diverge.

The difficult cases simulated at the end of this section seem to show robustness and good behavior even in multivariable or nonminimum phase cases. Yet, at the present time, it is felt that this was not rigorously proven and requires more research.

E. EXAMPLES

The numerous examples used for simulation may show the smoothness of this control system design method. Again, all adaptive gains were initially zero and no prior information was used for implementation, besides the assumption that some simple fitted feedforward configurations exist, pending the amount of a priori information.

Low-frequency sinusoidal waves and constant signals are used as output disturbances, and their effect is entirely felt at the plant output. No filtering (which could be used in practice) is used here, since the following simulations are only selected to show robustness and boundedness of all values involved in the adaptation process.

Example 3

The example used in [21], although stable and minimum phase, seems to challenge almost all adaptive control methods, especially in the presence of large input commands or disturbances.

The transfer function of the plant is

$$\frac{y_p(s)}{u_p(s)} = \frac{2}{s+1}\,\frac{229}{s^2 + 30s + 229}, \tag{111}$$

and the output of the plant is required to follow the output of the following first-order model:

$$\frac{y_m(s)}{u_m(s)} = \frac{1}{1 + s/3}. \tag{112}$$

A fixed feedforward configuration is heuristically selected to be

$$\dot{x}_s = u_s, \tag{113}$$

$$y_s = x_s. \tag{114}$$

The adaptation coefficients are $\sigma = 0.01$, $T = \bar{T} = 10{,}000I_3$, and $u_s = -100x_s + 5u_p$, and we only calculate the control (37)-(42).

The sinusoidal output disturbance is used to check robustness:

$$d_o = 0.08 \sin 10t. \tag{115}$$

Simulation results represented in Figs. 5a-d show the behavior of the adaptive system in different cases. The simulation starts

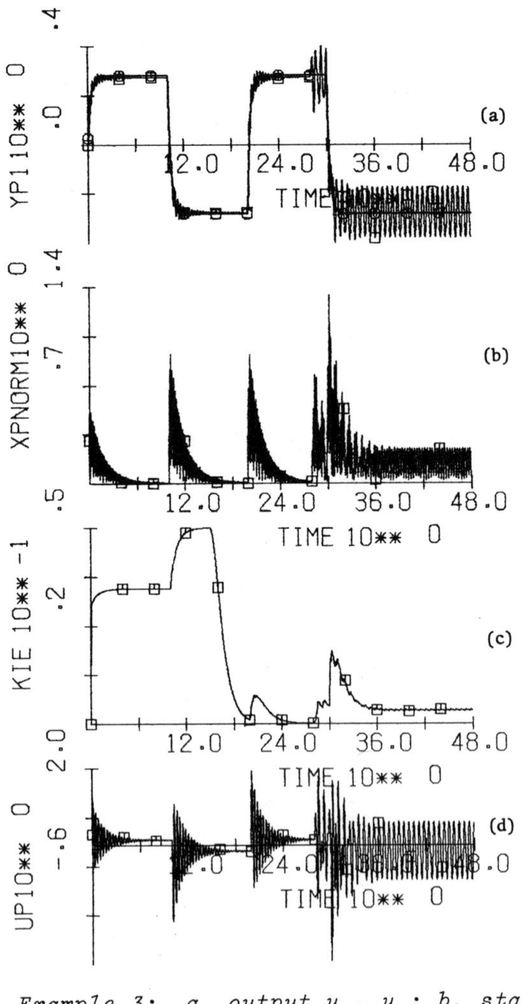

Fig. 5. Example 3: a, output y_m, y_p; b, state vector norm; c, gain K_{Ie}; d, control u_p.

with $\sigma = 0$ in a clean environment and shows good model following, with a small bias, due to the supplementary feedforward. However, the integral gain $K_{Ie}(t)$ tends to increase whenever the output tracking error is not zero. At $t = 15$ sec we use $\sigma = 0.01$, and the value of the integral gain decreases considerably with no apparent effect on the performance. At $t = 28$ sec the output disturbance is switched in and its effect is seen on the output of the plant and on the integral gain, which increases to some extent. No filtering (which can be used in practice) is used here, since the simulations are only intended to show robustness with disturbances.

Example 4

The two-input, two-output, four-state reactor used in [39] for multivariable linear system design is required here to track the output of the very simple model

$$\dot{x}_m(t) = \begin{bmatrix} -25 & 0 \\ 0 & -25 \end{bmatrix} x_m(t) + \begin{bmatrix} 0 \\ 1 \end{bmatrix} u_m(t), \tag{116}$$

$$y_m(t) = \begin{bmatrix} 1 & 0 \\ 0 & 1 \end{bmatrix} x_m(t) \tag{117}$$

for square-wave input commands.

The system matrices of the reactor are

$$A_p = \begin{bmatrix} 1.38 & -0.2047 & 6.715 & -5.676 \\ -0.5814 & -4.29 & 0 & 0.675 \\ 1.067 & 4.273 & -6.654 & 5.893 \\ 0.048 & 4.273 & 1.343 & -2.104 \end{bmatrix}, \tag{118}$$

$$B_p = \begin{bmatrix} 0 & 0 \\ 5.679 & 0 \\ 1.136 & -3.146 \\ 1.136 & 0 \end{bmatrix}, \quad C_p = \begin{bmatrix} 1 & 0 & 1 & -1 \\ 0 & 1 & 0 & 0 \end{bmatrix}. \tag{119}$$

The supplementary feedforward is selected simply as

$$\begin{bmatrix} \dot{x}_{s_1}(t) \\ \dot{x}_{s_2}(t) \end{bmatrix} = \begin{bmatrix} u_{s_1}(t) \\ u_{s_2}(t) \end{bmatrix}, \tag{120}$$

$$y_s(t) = x_s(t). \tag{121}$$

The adaptation coefficients are $\sigma = 0.02$, $T = \bar{T} = 1000I_8$, and $T_s = \bar{T}_s = 1000I_2$, while $r_s(t) \equiv x_s(t)$ and $K_{u_p} = I_2$.

In order to check the robustness of the adaptive system, large input and output disturbances were introduced as follows:

$$d_i(t) = \begin{bmatrix} 5 \\ 10 \sin 6t \end{bmatrix}, \quad d_o(t) = \begin{bmatrix} 0.03 \\ 0.04 \sin 10t \end{bmatrix}. \tag{122}$$

The results of simulations presented in Figs. 6a-d show robustness and small tracking errors.

Example 5

Another example is a two-input, two-output, eight-state missile [40] with rapidly changing parameters.

The plant matrices are

$$A = \begin{bmatrix} 3.23 & p & -476.0 & 0 & 228.0 & 0 & 0 & 0 \\ p & -3.23 & 0 & 476.0 & 0 & -228.0 & 0 & 0 \\ 0.39 & 0 & -1.93 & -0.8p & -415.0 & 0 & 0 & 0 \\ 0 & -0.39 & 0.8p & -1.93 & 0 & -415.0 & 0 & 0 \\ 0 & 0 & 0 & 0 & 0 & 0 & 75.0 & 0 \\ 0 & 0 & 0 & 0 & 0 & 0 & 0 & -75.0 \\ 0 & 0 & 22.4 & 0 & -300.00 & 0 & -150.0 & 0 \\ 0 & 0 & 0 & -22.4 & 0 & 300.0 & 0 & -150.0 \end{bmatrix}, \tag{123}$$

$$B^T = \begin{bmatrix} 0 & 0 & 0 & 0 & 0 & 0 & -1.0 & 0 \\ 0 & 0 & 0 & 0 & 0 & 0 & 0 & -1.0 \end{bmatrix}, \tag{124}$$

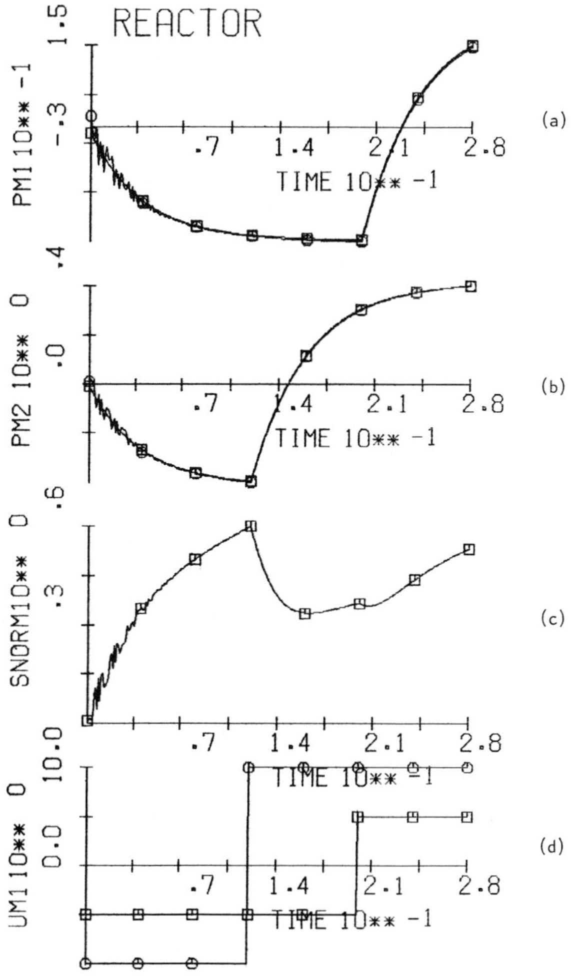

Fig. 6. Example 4: a, output y_{m_1}, y_{p_1}; *b, output* y_{m_2}, y_{p_2}; *c, state vector norm; d, input* u_{m_1}, u_{m_2}.

$$C = \begin{bmatrix} -2.99 & 0 & -1.19 & 0.123p & -27.64 & 0 & 0 & 0 \\ 0 & -2.99 & 0.123p & 1.19 & 0 & 27.64 & 0 & 0 \end{bmatrix}. \tag{125}$$

Besides the plant all other elements involved in the adaptation process (model reference, gains, coefficients) are the same as in the previous example. Note that some system parameters are functions of the roll rate p. This fact does not affect the adaptive controller design. The simulation starts with p = 10, then at t = 0.35 sec it replaces it (abruptly) by p = 20, and at t = 0.65 sec by p = 0. The adaptation coefficients are

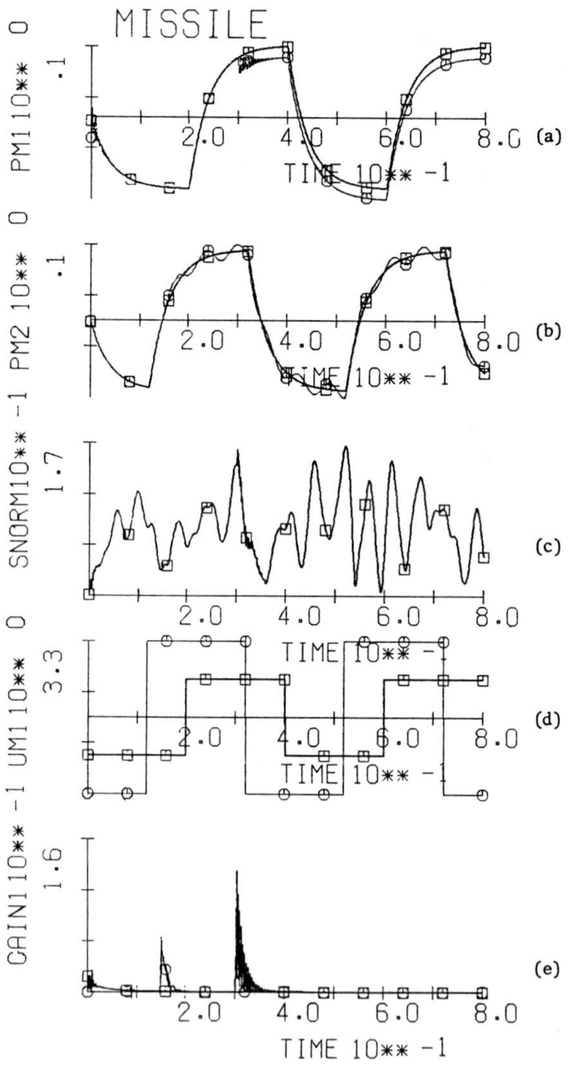

Fig. 7. Example 5: a, output y_{m_1}, y_{p_1}; b, output y_{m_2}, y_{p_2}; c, state vector norm; d, input u_{m_1}, u_{m_2}; e, gains $K_{e_{11}}$, $K_{e_{22}}$.

are $T = \bar{T} = 1000I_6$ and $T_s = \bar{T}_s = 1000I_4$, and all gains are computed through (91)-(102).

Results of simulation presented in Figs. 7a-e for a realistic disturbed environment show good output tracking and robustness. The input disturbances are introduced from the start and their effect is negligible. The large output disturbances are introduced at t = 0.35 and at t = 0.45 sec and, correspondingly, create output tracking errors; however, all values are bounded and the system behaves quite well. The changes in the parameter p are felt mainly by the adaptive gains—they may change to adapt to the new situation—and may hardly have any effect on the performance of the adaptive control system.

Example 6

The sixth example is, again, the unstable and nonminimum phase oscillatory representation of a helicopter [38]. The plant is required to follow the behavior of the first-order model (64). Since it is assumed that the prior knowledge used in the first part of the chapter is unknown, the heuristic feedforward configuration is again selected to be (113)-(114), with $r_s(t) \equiv x_s(t)$ and $K_{u_p} = 0.1$.

The adaptation coefficients are $T = \bar{T} = 1000I_3$ and $T_s = \bar{T}_s = 1000$, and the output disturbance is $d_o = 0.01 \sin 10t$.

Note that high coefficients do not mean high gains in the context of the present algorithm.

The simulation was run first with $\sigma = 0.0$ for 1 sec, and the results presented in Figs. 8a-c seem very discouraging, especially with respect to gain divergence. However, changing the parameter value to $\sigma = 0.001$ at t = 1 sec results in a robust and generally well-behaving control system.

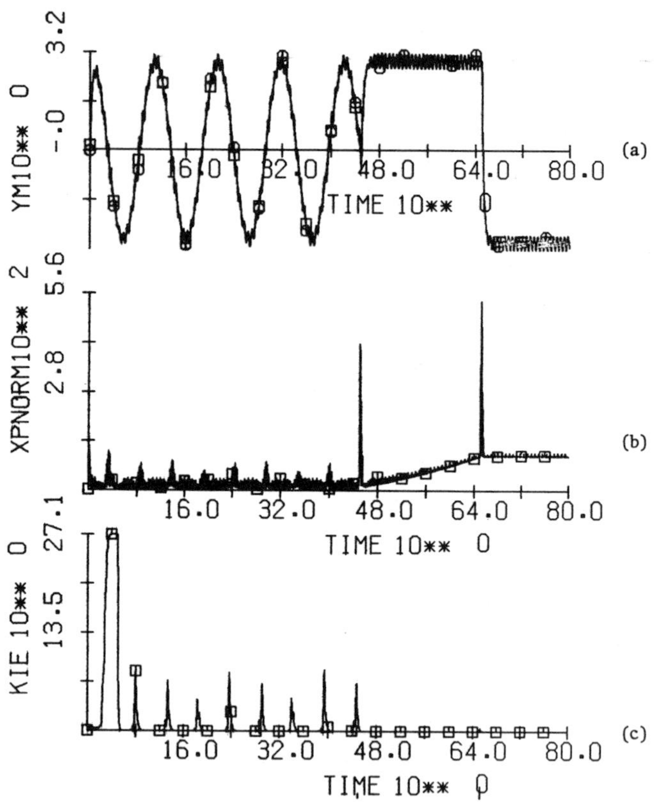

Fig. 8. Example 6: a, output y_m, y_p; b, state vector norm; c, gain K_{Ie}.

Example 7

The last example may illustrate a possible role for adaptive control design methods as a complement of linear control design.

The selected controlled plant is a classical model for missile autopilot design [41]. A schematic representation of the plant is shown in Fig. 9. The aim of the adaptive controller is to establish the acceleration response of the missile, while considering that the velocity u varies between 250 m/sec and 750 m/sec, and that the other parameters are $u_0 = 500$, $y_z = 180(u/u_0)^2$, $n_z = -500(u/u_0)^2$, $n_r = -3(u/u_0)^2$, $y_n = -3(u/u_0)$, $n_n = u/u_0$, $k_s = 0.007$, $\omega_{n_s} = 180$, $\mu_s = 0.5$, $k_g = 30$, and $c = 0.5$

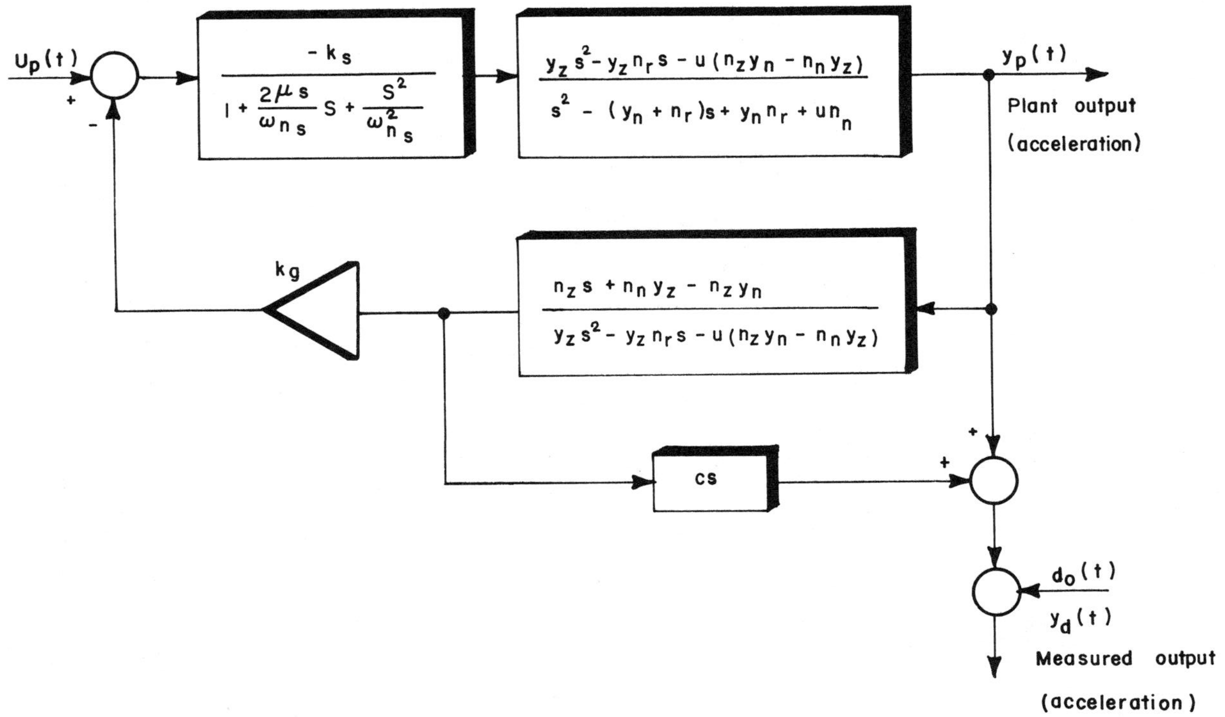

Fig. 9. Schematic representation of lateral missile control system.

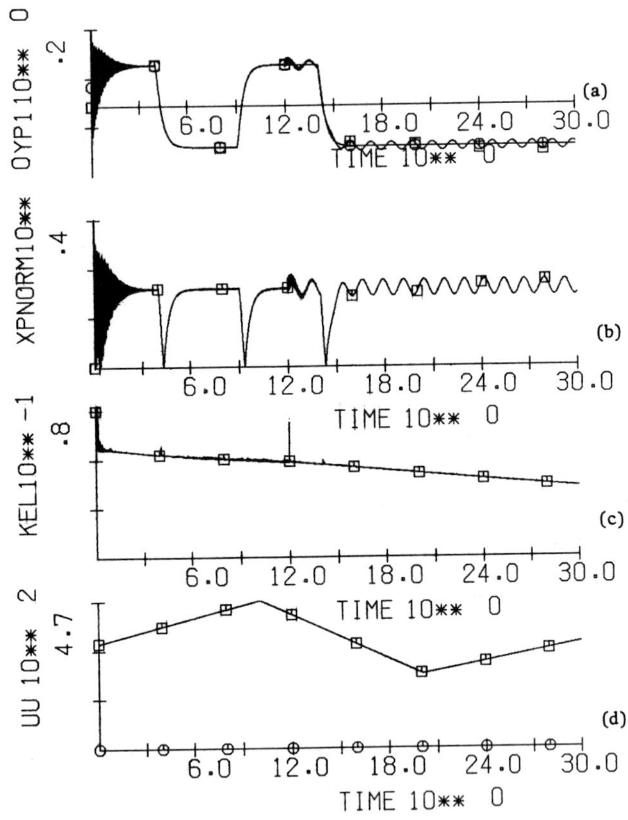

Fig. 10. Example 7: a, output y_m, y_p; b, state vector norm; c, gain K_{Ie}; d, velocity u(t).

The missile is required to follow the input-output behavior of the model (64) and the supplementary configuration is again (113)-(114). The adaptation coefficients are $T = \bar{T} = 1000I_3$, $T_s = \bar{T}_s = 1000I_2$, and $\sigma = 0.001$.

The results of simulations are represented in Figs. 10a-d with linear variation of the velocity u between the limits. After almost perfect tracking, the (large) output disturbance $d_o = 0.03 \sin 5t$ is introduced and the adaptive system shows robustness and boundedness of all values.

The examples of this section presented different controllers whose complexity of implementation was a function of the amount of prior knowledge given for each case.

The reader with practical applications in his mind may very well find out that the simpler adaptive controller configurations may be the most useful, since in practical situations, the representations of planes, structures, missiles, or processes are not totally unknown, as was sometimes assumed here.

V. CONCLUSIONS

In this article a simple and robust adaptive control algorithm for multivariable continuous-time systems was first presented and global stability and robustness with respect to boundedness in the presence of parasitic input and output disturbances was proven. The proposed algorithm is applicable only in systems that can be stabilized via output feedback. A constant feedforward configuration must be selected a priori for implementation of the algorithm.

Modified adaptive control procedures were successively presented. These procedures need prior knowledge of the minimal order of a stabilizing feedback, which is equivalent to the minimal order of the feedforward that can guarantee satisfaction of the positivity conditions.

The order of the plant and its pole excess are unknown and not needed for implementation. Under these assumptions, the adaptive control system is robust with respect to parasitic dynamics and disturbances.

These properties together with the simplicity of implementation compared with other adaptive control algorithms may make the procedures presented in this article a useful step toward implementation of adaptive control in realistic environments and in multivariable large-scale systems.

Extension of these techniques to discrete-time adaptive control systems [42] and further reduction of the prior knowledge needed for implementation are the subject of present research.

APPENDIX A: THE DERIVATIVE OF THE LYAPUNOV FUNCTION (49)

Let

$$V = V_1(t) + V_2(t), \tag{A.1}$$

where

$$V_1(t) = e_x(t) P e_x(t), \tag{A.2}$$

$$V_2(t) = \operatorname{tr}\left[\left(K_I(t) - \tilde{K}\right) T^{-1} \left(K_I(t) - \tilde{K}\right)^T\right], \tag{A.3}$$

$$\dot{V}_1(t) = \dot{e}_x^T(t) P e_x(t) + e_x^T(t) P \dot{e}_x(t). \tag{A.4}$$

Substituting $\dot{e}_x(t)$ from (44) gives

$$\dot{V}_1(t) = e_x^T(t)(PA_c + A_c^T P) e_x(t) - 2e_x^T(t) PB[K(t) - \tilde{K}]r(t)$$
$$- 2e_x^T(t) PF(t), \tag{A.5}$$

and

$$\dot{V}_2(t) = 2 \operatorname{tr}\left[\left(K_I(t) - \tilde{K}\right) T^{-1} \dot{K}_I^T(t)\right], \tag{A.6}$$

$$\dot{V}_2(t) = 2 \operatorname{tr}\left[\left(\tilde{K}_I(t) - \tilde{K}\right)\left(e_y(t) r^T(t) - \sigma K_I(t)\right)^T\right]. \tag{A.7}$$

Substituting in the second term

$$K_I(t) = K(t) - K_p(t) = K(t) - e_y(t) r^T(t) \overline{T} \tag{A.8}$$

gives

$$\dot{V}_2(t) = -2\sigma \ \mathrm{tr}\left[\left(K_I(t) - \tilde{K}\right)K_I^T(t)\right] + 2e_y^T(t)[K(t) - \tilde{K}]r(t)$$

$$- 2e_y^T(t)e_y(t)r^T(t)\bar{T}r(t). \tag{A.9}$$

By using (34)-(35) we obtain from (A.5)

$$\dot{V}_1(t) = -e_x^T(t)Qe_x(t) - 2e_x^T(t)C^T[K(t) - \tilde{K}]r(t)$$

$$- 2e_x^T(t)PF(t), \tag{A.10}$$

and by substituting $e_y(t)$ from (36) we obtain from (A.9)

$$\dot{V}_2(t) = -2\sigma \ \mathrm{tr}\left[\left(K_I(t) - \tilde{K}\right)\left(K_I(t) - \tilde{K}\right)^T\right]$$

$$+ 2\sigma \ \mathrm{tr}\left[\left(K_I(t) - \tilde{K}\right)\tilde{K}^T\right]$$

$$+ 2e_x^T(t)C^T[K(t) - \tilde{K}]r(t) - d_o^T(t)[K(t) - \tilde{K}]r(t). \tag{A.11}$$

Adding (A.10) and (A.11) gives (50).

APPENDIX B: THE DERIVATIVE OF THE LYAPUNOV FUNCTION (109)

Differentiation of V(t) in (109) gives

$$\dot{V}(t) = e_x^T(t)P\dot{e}_x(t) + \dot{e}_x^T(t)Pe_x(t)$$

$$+ 2 \ \mathrm{tr}\left[\left(K_I(t) - \tilde{K}\right)T^{-1}\dot{K}_I^T(t)\right]$$

$$+ 2 \ \mathrm{tr}\left[\left(K_{sI}(t) - \tilde{K}_s\right)T_s^{-1}\dot{K}_{sI}^T(t)\right]. \tag{B.1}$$

Substituting the corresponding derivatives gives

$$\dot{V}(t) = e_x^T(t)(PA + A^TP)e_x(t) - 2e_x^T(t)PB[K(t) - \tilde{K}]r(t)$$

$$- 2e_x^T(t)PF(t) + 2e_x^T\begin{bmatrix} P_{11} & P_{12} \\ P_{21} & P_{22} \end{bmatrix}\begin{bmatrix} 0 \\ B_s \end{bmatrix}\left[K_s(t) - \tilde{K}_s\right]x_s(t)$$

$$+ 2\ \mathrm{tr}\left[\left(K_I(t) - \tilde{K}\right)T^{-1}T\left(e_y r^T(t) - \sigma K_I(t)\right)^T\right]$$

$$+ 2\ \mathrm{tr}\left[\left(K_{sI} - \tilde{K}_s\right)T_s^{-1}T_s\left(e_{y_s}(t)x_s^T(t) - \sigma K_{sI}(t)\right)^T\right]. \tag{B.2}$$

By using (76)-(77) and (108), we obtain after some algebraic manipulations

$$\begin{aligned}
\dot{V}(t) = {} & e_x^T(t)Qe_x(t) - 2e_x^T(t)C[K(t) - \tilde{K}]r(t) - 2e_x^T(t)PF(t) \\
& + 2e_p^T(t)P_{12}B_s\left[K_s(t) - \tilde{K}_s\right]r_s(t) \\
& + 2e_s^T(t)P_{22}B_s\left[K_s(t) - \tilde{K}_s\right]r_s(t) \\
& - 2\sigma\ \mathrm{tr}\left[\left(K_I(t) - \tilde{K}\right)\left(K_I - \tilde{K}\right)^T\right] - 2\sigma\ \mathrm{tr}\left[\left(K_I(t) - \tilde{K}\right)\tilde{K}^T\right] \\
& - 2e_y^T(t)e_y(t)r^T(t)\overline{T}r(t) + 2e_x^T(t)C[K(t) - \tilde{K}]r(t) \\
& - d_o^T(t)[K(t) - \tilde{K}]r(t) \\
& - 2\sigma\ \mathrm{tr}\left[\left(K_{sI}(t) - \tilde{K}_s\right)\left(K_{sI}(t) - \tilde{K}_s\right)^T\right] \\
& + 2\sigma\ \mathrm{tr}\left[\left(K_{sI}(t) - \tilde{K}_s\right)\tilde{K}_s^T\right] - 2e_s^T(t)C_s^T\left[K_s(t) - \tilde{K}_s\right]r_s(t) \\
& - 2y_s^T(t)y_s(t)r_s^T(t)\overline{T}_s r_s(t).
\end{aligned} \tag{B.3}$$

Rearranging the terms finally gives (110).

REFERENCES

1. K. J. ASTRÖM and B. WITTENMARK, "On Self-Tuning Regulators," *Automatica 9*, 185-199, (1973).

2. K. J. ASTRÖM, U. BORISSON, L. LJUNG, and B. WITTENMARK, "Theory and Applications of Self-Tuning Regulators," *Automatica 13*, 457-476 (1977).

3. D. W. CLARKE and P. J. GAWTHROP, "Self-Tuning Controller," *Proc. IEE-D 122*, 929-934 (1975).

4. P. E. WELLSTEAD, J. M. EDMUNDS, D. PRAGER, and P. ZANKER, "Self-Tuning Pole/Zero Assignment Regulators," *Int. J. Control 30*, 1-26 (1979).

5. P. C. PARKS, "Lyapunov Redesign of Model Reference Adaptive Control Systems," *IEEE Trans. Autom. Control AC-11*, 362-367 (1966).

6. I. D. LANDAU, "A Survey of Model-Reference Adaptive Techniques: Theory and Applications," *Automatica 10*, 353-360 (1974).

7. R. V. MONOPOLI, "Model Reference Adaptive Control with an Augmented Error Signal," *IEEE Trans. Autom. Control AC-19*, 474-484 (1974).

8. K. S. NARENDRA and L. S. VALAVANI, "Direct and Indirect Adaptive Control," *Automatica 15*, 653-663 (1979).

9. M. BALAS and C. R. JOHNSON, JR., "Adaptive Identification and Control of Large-Scale or Distributed Parameter Systems Using Reduced-Order Models," *Workshop Appl. Adapt. Control, 2nd, Yale University, New Haven, Connecticut*, 175-181 (1981).

10. B. EGARDT, "Stability of Adaptive Controllers, Lecture Notes in Control and Information Sciences," Vol. 20, Springer-Verlag, Berlin and New York, 1979.

11. H. ELLIOTT, "Direct Adaptive Pole Placement with Application to Nonminimum Phase Systems," *IEEE Trans. Autom. Control AC-27*, 72-726 (1982).

12. A FEUER and A. S. MORSE, "Adaptive Control of Single-Input, Single-Output Linear Systems," *IEEE Trans. Autom. Control AC-23*, 557-569, August 1978.

13. G. C. GOODWIN, P. J. RAMADGE, and P. E. GAINES, "Discrete-Time Multivariable Adaptive Control," *IEEE Trans. Autom. Control AC-25*, 449-456 (1980).

14. T. LONESCU and R. V. MONOPOLI, "Discrete Model Reference Adaptive Control with an Augmented Error Signal," *Automatica 13*, 507-517 (1977).

15. I. D. LANDAU, "Adaptive Control—The Model Reference Approach," Dekker, New York, 1979.

16. L. LJUNG, "On Positive Real Transfer Functions and the Convergence of Some Recursive Schemes," *IEEE Trans. Autom. Control AC-22*, 539-551 (1977).

17. A. S. MORSE, "Global Stability of Parameter's Adaptive Control Systems," *IEEE Trans. Autom. Control AC-25*, 433-439 (1980).

18. K. S. NARENDRA, Y. H. LIN, and L. S. VALAVANI, "Stable Adaptive Controller Design—Part II: Proof of Stability," *IEEE Trans. Autom. Control AC-25*, 440-448 (1980).

19. K. J. ASTRÖM, "Theory and Applications of Adaptive Control—A Survey," *Automatica 19*, No. 5, 471-481 (1983).

20. K. S. NARENDRA and R. V. MONOPOLI (ed.), "Applications of Adaptive Control," Academic Press, New York, 1980.

21. C. ROHRS, L. VALAVANI, M. ATHANS, and G. STEIN, "Stability Problems of Adaptive Control Algorithms in the Presence of Unmodelled Dynamics," *Proc. 21st IEEE Conf. Decision Control, Orlando 1*, 3-11 (1983).

22. J. BROUSSARD and P. BERRY, "Command Generator Tracking—The Continuous-Time Case," Technical Report TIM-612-1, TASC (1978).

23. K. SOBEL, H. KAUFMAN, and L. MABIOUS, "Implicit Adaptive Control for a Class of MIMO Systems," *IEEE Trans. Aerosp. Electron. Syst. AES-18*, 576-590 (1982).

24. K. SOBEL and H. KAUFMAN, "Direct Model Reference Adaptive Control for a Class of MIMO Systems," this volume.

25. I. BAR-KANA, "On (Some) Rehabilitation of Positive Realness in Adaptive Control," submitted, (1987).

26. I. BAR-KANA, "Direct Multivariable Adaptive Control with Application to Large Structural Systems," Ph.D. Thesis, Rensselaer Polytechnic Institute, Troy, New York, 1983.

27. I. BAR-KANA and H. KAUFMAN, "Global Stability and Performance of a Control," *Int. J. Control 42*, 1491-1505 (1985).

28. I. BAR-KANA, H. KAUFMAN, and M. BALAS, "Model Reference Adaptive Control of Large Structural Systems," *AIAA J. Guidance Dyn. Control 6*, 112-118 (1983).

29. I. BAR-KANA and H. KAUFMAN, "Some Applications of Direct Adaptive Control to Large Structural Systems," *AIAA J. Guidance Dyn. Control 7*, 717-724 (1984).

30. S. J. WANG, C. H. C. IH, Y. H. LIN, and E. M. METTLER, "Space Station Dynamic Modeling, Disturbance Accomodation, and Adaptive Control," *Workshop Identif. Control Flexible Space Struct., San Diego, California* (1984).

31. C. H. C. IH., S. J. WANG, and C. LEONDES, "An Investigation of Adaptive Control Techniques for Space Stations," *Proc. ACC, 1985, Boston, Massachusetts* (1985).

32. P. IOANNOU and P. V. KOKOTOVIC, "Singular Perturbations and Robust Redesign of Adaptive Control," *Proc. 21st IEEE Conf. Decision Control, Orlando*, 24-29 (1982).

33. I. BAR-KANA and H. KAUFMAN, "Direct Adaptive Control with Bounded Tracking Errors," *Proc. 22nd IEEE Conf. Decision Control, San Antonio, Texas*, December 1983.

34. I. BAR-KANA and H. KAUFMAN, "Robust Simplified Adaptive Control for a Class of Multivariable Systems," *Proc. 24th IEEE Conf. Decision Control, Fort Lauderdale, Florida*, December 1985, pp. 141-146.

35. I. BAR-KANA and H. KAUFMAN, "Discrete-Direct Multivariable Adaptive Control," *Proc. 1st IFAC Workshop Adapt. Syst. Control Signal Process., San Francisco, California* (1983).

36. B. D. O. ANDERSON and S. VONGPANITLERD, "Network Analysis and Synthesis," Prentice-Hall, Englewood Cliffs, New Jersey, 1973.

37. R. J. BENHABIB, "Stability of Large Space Structure Control Systems Using Positivity Concepts," *J. Guidance Control 5*, 487-494, September-October 1981.

38. H. KURZ, R. ISERMAN, and R. SCHUMANN, "Experimental Comparison and Application of Various Parameter-Adaptive Control Algorithms," *Automatica 16*, 117-130 (1980).

39. A. G. MACFARLANE and B. KOUVRITAKIS, "A Design Technique for Linear Multivariable Feedback Systems," *Int. J. Control 25*, 837-874 (1977).

40. J. M. EDMUNDS, "Control System Design and Analysis Using Closed-Loop Nyquist and Bode Arrays," *Int. J. Control 16*, 117-130 (1980).

41. P. GARNELL and D. J. EAST, "Guided Weapon Control Systems," Pergamon, Oxford, 1977.

42. I. BAR-KANA, "On Positive Realness in Discrete-Time Adaptive Control Systems," *Int. J. Syst. Sci. 17*, 1001-1006 (1986); also *Proc. ACC, Seattle, Washington*, 1440-1443 (1986).

Discrete Averaging Principles and Robust Adaptive Identification

ROBERT R. BITMEAD

Department of Systems Engineering
Research School of Physical Sciences
Australian National University
Canberra ACT 2601, Australia

C. RICHARD JOHNSON JR.

School of Electrical Engineering
Cornell University
Ithaca, New York 14853

I. INTRODUCTION AND OUTLINE

We present here an application of the theory of averaging to the analysis of the robustness of adaptive identification methods, focusing for clarity on the scalar gain or "gradient-based" output error scheme. The aim is to derive broad design principles for the achievement of this robustness. The key requirements developed are for slow adaptation relative to the plant dynamics, persistently exciting signals being used in the identification, and local stability and positivity properties of certain parameter values and signals. The relationship of these requirements to free design variables of adaptive identification is discussed.

The structure and flow of this article are as follows. In Section II we discuss, in moderate detail, the setting for our problem, and, in particular, focus on the distinctions in formulation between identification, adaptive identification, and robust adaptive identification. Section III is devoted to a presentation of known results on the equation error and output error adaptive identification algorithms, which serve as our candidate representative algorithms. Difference equations describing the propagation of parameter errors are derived. In Section IV we develop discrete averaging theorems that allow a stability analysis of certain difference equations depending on a small parameter. These averaging results permit the qualitative analysis of the stability of certain nonlinear, time-varying difference equations through the study of related linear, time-invariant equations and are directed toward the error equations of the previous section. Section V contains the explicit matching of the error models with the averaging results and discusses some connections with previous results. Finally, in Section VI we discuss the implications of the article's results as foreshadowed above and make some suggestions concerning algorithm robustness achievement and enhancement.

II. IDENTIFICATION, ADAPTIVE IDENTIFICATION, AND ROBUST ADAPTIVE IDENTIFICATION

In presenting a cogent theory of robust adaptive identification with the intent of producing sound guidelines for practical implementation, the need for adequate development of the problem formulation is manifest. Our aim in this section will be somewhat taxonomic, somewhat philosophical, and somewhat technical as we attempt to specify and justify our particular version of a mythological figment—*the* robust adaptive identification problem. In spite of these Platonistic overtones, we shall develop, in turn, in this section, settings for each of the topics of the title in order that the validity of perceived implications becomes apparent.

The notion of system identification is that of modeling the causal relationship of one (output) time series $\{y_k\}$ to another (input) time series $\{u_k\}$ using experimental, measured data. Aspects of such model fitting have been dealt with by many authors [1,2], with the key feature being to fit a model that optimizes some criterion of fit over some class of admissible models, given the existing data. The particular class that we shall consider here is the set of scalar, linear, fixed-degree, deterministic autoregressive moving-average (ARMA) models.*

$$y_k = -a_1 y_{k-1} - a_2 y_{k-2} - \cdots - a_n y_{k-n} + b_1 u_{k-1} + b_2 u_{k-2} + \cdots + b_n u_{k-n}, \tag{1}$$

where the free parameters are the coefficients $\{a_1, \ldots, a_n, b_1, \ldots, b_n\}$. These parameters define polynomials in the delay operator z as follows:

$$A(z) = 1 + a_1 z + a_2 z^2 + \cdots + a_n z^n,$$

$$B(z) = b_1 z + b_2 z^2 + \cdots + b_n z^n.$$

Throughout the article we shall assume that the true plant is bounded-input/bounded-output stable so that bounded $\{u_k\}$ implies bounded $\{y_k\}$. When (1) exactly describes the plant, this stability corresponds to A(z) having no zeros inside the unit circle.

In performing a particular system identification it is apparent that the prevailing experimental conditions, that is,

*In the statistical literature these models would be named ARX since $\{u_k\}$ is measurable here. However, we shall stick with the historical engineering nomenclature.

the initial conditions and values taken by $\{u_k\}$, play a major role in determining an optimizing solution [2]. This dependence on experimental conditions could be severe in instances where, say, the two time series are related in a nonlinear fashion—the ARMA model then describing the linearized relationship about an operating point, at best—but would be less crucial in cases where $\{u_k\}$ and $\{y_k\}$ truly are related by such an ARMA description as (1). Indeed, it is a nontrivial matter to guarantee the existence of a limiting, optimizing parameter value as the number of data points becomes infinite, the usual techniques requiring stationarity assumptions either on the data $\{y_k\}$, $\{u_k\}$ or on the existence of an exact, fixed model within the admissible class.

It is in respect of this latter point that adaptive identification arises as a modified objective where, even though optimizing parameter values may be derived for any finite number of data points, the validity of any one model for the complete time history of the processes is seen as unrealistic and one attempts to fit models whose horizon of validity is more restricted. In particular, it is frequently stated that the goal of adaptive identification is to track slow time variations of systems, and it is easy to interpret this, for example, in terms of tracking linearized models as the operating point alters slowly. One hallmark of adaptive identification is that one typically does not presuppose any explicit knowledge of the time variation of such systems other than knowledge of time scale but modifies the parameter estimation procedure to enhance local (in time) validity of the model at the cost of consistency of limiting parameter values. Adaptive identification schemes are usually recursive with nonvanishing step sizes and are often real-time processes.

If one has available a description of the variation of the best-fitting model, then it is again possible to extend the parameter estimation scheme to take this into account explicitly. The more usual notion, however, is that an a priori design compromise is made trading-off consistency in stationary infinite-horizon problems against the ability to follow slow time variations of the (possibly hypothetical) best-fitting model over finite time windows. The distinction that we draw between identification and adaptive identification, therefore, lies in our presumptions concerning the time scale of validity of models and in the consequent admissions of variability and effect of the experimental conditions or of the underlying relationship between the time series.

If consistency (or convergence of finite-time optimizing or suboptimal parameter estimates to the infinite-time optimizing value) is taken as a yardstick of identifier performance, we may ask, "What is the performance measure for an adaptive parameter estimator?" Since we have been deliberately vague and nebulous in characterizing time variations to be tracked as, indeed, we have been concerning the true nature of the relationship between the time series, our performance measure for adaptive identification is similarly qualitative, with no precise notion of explicit optimization of this measure implied. In particular, we ask that, for $\{y_k\}$, $\{u_k\}$ either related exactly by an ARMA model (1) or only approximately related by a best model at some time, the parameter estimate should be "close" to the exact or best value. Closeness here could be measured in terms of parameter vector deviation norm, prediction error values, frequency response of related linear time-invariant systems, etc. We shall adopt the former measure since it will imply the others in general.

Robust adaptive identification is concerned with achieving closeness of the parameter estimate to the nominal theoretically best value at any particular time in spite of inexactness in the modeling, time variations of the best value, etc. The ability to specify theoretically this nominal parameter value limits the attainable closeness. However, there is a strong principle in this theory that design choices should be able to be made that allow improved closeness as the deviation diminishes from the ideal situation, where a single fixed, limiting model optimizes the fit. Thus an identification scheme that arbitrarily assigned a bounded parameter value would not be considered robust in this context. These points are discussed in [3].

Having outlined our aims in robust adaptive identification in this section, together with some philosophical deliberations, we move on in Section III to present the particular recursive adaptive ARMA identification algorithm for equation error and output error methods. Our aim is to establish robustness conditions for these standard algorithms rather than to derive new procedures intended to improve this, as yet poorly understood, robustness. We shall derive error equations describing the deviation of parameter estimates for these schemes from nominal values and discuss the stability properties of these equations. In Section V robustness will then be associated with bounded-input/bounded-output stability of these equations, which in turn will be associated with exponential asymptotic stability of the homogeneous error equations. Averaging theory is brought to bear to establish these latter results.

III. ADAPTIVE IDENTIFICATION ALGORITHMS AND ERROR SYSTEMS

In line with our choice of parametrized linear ARMA models, we shall consider in this section standard recursive equation error and output error adaptive estimation schemes. Our development here follows closely that of Goodwin and Sin [4], whom we refer to for several, now standard, results. We shall specialize to gradient-based estimation algorithms and indicate later possible generalizations to least squares or Newton schemes.

We write the ARMA model (1) as

$$y_k = \phi_{k-1}^T\theta, \tag{2}$$

where the information vector ϕ_{k-1} is given by

$$\phi_{k-1} = (y_{k-1}, y_{k-2}, \ldots, y_{k-n}, u_{k-1}, \ldots, u_{k-n})^T \tag{3}$$

and the parameter vector θ by

$$\theta = (-a_1, -a_2, \ldots, -a_n, b_1, \ldots, b_n)^T. \tag{4}$$

The standard equation error adaptive parameter estimator is then

$$\hat{\theta}_k = \hat{\theta}_{k-1} + \frac{1}{c + \phi_{k-1}^T\phi_{k-1}}\,\phi_{k-1}^T\left(y_k - \phi_{k-1}^T\hat{\theta}_{k-1}\right), \tag{5}$$

where $\hat{\theta}_k$ is the vector of parameter estimates at time k and $c > 0$. The coefficient c dictates the step size of the parameter update algorith. Writing $\tilde{\theta}_k = \theta - \hat{\theta}_k$, we may display (5) in an alternative (noncausal) form as in [4, Section 3.9]:

$$\begin{aligned}\tilde{\theta}_k &= \tilde{\theta}_{k-1} - \frac{1}{c}\,\phi_{k-1}\phi_{k-1}^T\tilde{\theta}_k \\ &= \tilde{\theta}_{k-1} + \frac{1}{c}\,\phi_{k-1}\left(y_k - \phi_{k-1}^T\tilde{\theta}_k\right).\end{aligned} \tag{6}$$

In the "ideal" case, where $\{y_k\}$ and $\{u_k\}$ exactly satisfy (2) for some θ, we have the following results from [4, Chapter 3] and [5].

Lemma 1 [4,5]

For the algorithm (5) with $\{y_k\}$ and $\{u_k\}$ exactly satisfying (2), we have

$$\lim_{k\to\infty}\left[c + \phi_{k-1}^T\phi_{k-1}\right]^{-1/2}\left[y_k - \phi_{k-1}\hat{\theta}_{k-1}^T\right] = 0, \tag{7}$$

so that, if $\{\phi_k\}$ is bounded,

$$\lim_{k\to\infty}\left[y_k - \phi_{k-1}^T\hat{\theta}_{k-1}\right] = 0. \tag{8}$$

Further, if $\{u_k\}$ is persistently exciting, that is, for all k there exists a fixed integer ℓ and fixed positive values ρ_1, ρ_2 such that

$$\rho_1 I \geq \sum_{i=k}^{k+\ell}\begin{bmatrix} u_{i+2n} \\ u_{i+2n-1} \\ \vdots \\ u_{i+1}\end{bmatrix}[u_{i+2n}, u_{i+2-1}, \ldots, u_{i+1}] \geq \rho_2 I, \tag{9}$$

and the plant polynomials A(z), B(z) defined by (4) are relatively prime, then $\left\{\hat{\theta}_k\right\}$ converges exponentially fast to θ.

An alternative parameter estimation algorithm of Landau [6] is the output error scheme, where two changes are made. First, one uses

$$\bar{\phi}_k^T = \left[\hat{y}_k, \hat{y}_{k-1}, \ldots, \hat{y}_{k-n+1}, u_k, \ldots, u_{k-n+1}\right], \tag{10}$$

where (the a priori model output)

$$\hat{y}_k = \bar{\phi}_{k-1}^T\hat{\theta}_{k-1}. \tag{11}$$

This "output error" alteration to equation error schemes has a beneficial effect on bias arising in ARMA parameter estimation in noise. The second modification is that one may attempt to mimic (6) using an a posteriori error

$$\eta_k = y_k - \bar{\phi}_{k-1}^T \hat{\theta}_k, \tag{12}$$

and, further, that a generalized a posteriori output error may be used:

$$\bar{\eta}_k = \eta_k + d_1\eta_{k-1} + \cdots + d_n\eta_{k-n} \triangleq D(z)\eta_k. \tag{13}$$

A causal method for implementing this scheme is

$$\hat{\theta}_k = \hat{\theta}_{k-1} + \frac{1}{c + \bar{\phi}_{k-1}^T \bar{\phi}_{k-1}} \bar{\phi}_{k-1}\nu_k, \tag{14}$$

$$\nu_k = y_k - \bar{\phi}_{k-1}^T \hat{\theta}_{k-1} + d_1\eta_{k-1} + \cdots + d_n\eta_{k-n}, \tag{15}$$

together with (12). We shall now derive the explicit noncausal error equation that mimics (6) for output error and leads directly to the appearance of a transfer function that classically has been assumed to be strictly positive real in order to prove stability. The noncausal equivalent derived in [4, Section 3.5] is

$$\hat{\theta}_k = \hat{\theta}_{k-1} + \frac{1}{c} \bar{\phi}_{k-1}\bar{\eta}_k, \tag{16}$$

which, in turn, admits the total description

$$\bar{\phi}_k = \begin{bmatrix} -\hat{a}_1 & -\hat{a}_2 & \cdots & -\hat{a}_n & \hat{b}_1 & \cdots & \hat{b}_n \\ 1 & 0 & \cdots & 0 & 0 & & 0 \\ \vdots & & & \vdots & \vdots & & \vdots \\ 0 & & 1 & 0 & 0 & & 0 \\ 0 & 0 & \cdots & 0 & 0 & \cdots & 0 \\ 0 & 0 & \cdots & 0 & 1 & \cdots & 0 \\ \vdots & \vdots & & \vdots & & & \vdots \\ 0 & 0 & & 0 & 0 & 1 & 0 \end{bmatrix} \bar{\phi}_{k-1} + \begin{bmatrix} 0 \\ 0 \\ \vdots \\ 0 \\ 1 \\ 0 \\ \vdots \\ 0 \end{bmatrix} u_k, \tag{17}$$

$$\hat{\theta}_k = \hat{\theta}_{k-1} + \frac{1}{c} \bar{\phi}_{k-1} D(z)\left[y_k - \bar{\phi}_{k-1}^T \hat{\theta}_k\right], \tag{18}$$

where the estimated values in the system matrix (17) are components of $\hat{\theta}_k$ and D(z) is the polynomial delay operator apearing in (13).

Following [4], we define two errors,

$$\tilde{\theta}_k = \theta - \hat{\theta}_k, \tag{19}$$

$$\tilde{\phi}_k = \phi_k - \overline{\phi}_k, \tag{20}$$

and write (17) as

$$\overline{\phi}_k = F\left(\hat{\theta}_k\right)\overline{\phi}_{k-1} + Gu_k, \tag{21}$$

$$\phi_k = F(\theta)\phi_{k-1} + Gu_k. \tag{22}$$

Subtracting yields

$$\tilde{\phi}_k = F(\theta)\tilde{\phi}_{k-1} + e_1\overline{\phi}_{k-1}\tilde{\theta}_k, \tag{23}$$

where $e_1^T = [1, 0, \ldots, 0]$, while we may write the a posteriori output error (12) as

$$y_k - \overline{\phi}_{k-1}^T\hat{\theta}_k = \overline{\phi}_{k-1}^T\tilde{\theta}_k + \tilde{\phi}_{k-1}\theta. \tag{24}$$

The algorithm (18) now becomes

$$\tilde{\phi}_k = F(\theta)\tilde{\phi}_{k-1} + e_1\overline{\phi}_{k-1}^T\tilde{\theta}_k, \tag{25}$$

$$\tilde{\theta}_k = \tilde{\theta}_{k-1} - \frac{1}{c}\overline{\phi}_{k-1}D(z)\left[\overline{\phi}_{k-1}^T\tilde{\theta}_k + \theta^T\tilde{\phi}_{k-1}\right], \tag{26}$$

or, alternatively,

$$\tilde{\phi}_k = F(\theta)\tilde{\phi}_{k-1} + e_1\phi_{k-1}^T\tilde{\theta}_k - e_1\tilde{\phi}_{k-1}^T\tilde{\theta}_k, \tag{27}$$

$$\tilde{\theta}_k = \tilde{\theta}_{k-1} - \frac{1}{c}\phi_{k-1}D(z)\left[\phi_{k-1}^T\tilde{\theta}_k - \tilde{\phi}_{k-1}^T\hat{\theta}_k\right]$$
$$+ \frac{1}{c}\tilde{\phi}_{k-1}D(z)\left[\phi_{k-1}^T\tilde{\theta}_k - \tilde{\phi}_{k-1}^T\hat{\theta}_k\right]. \tag{28}$$

Using the former description, we may write from (25)

$$\tilde{\phi}_{k-1} = [z^{-1}I - F(\theta)]^{-1}e_1\overline{\phi}_{k-1}^T\tilde{\theta}_k, \tag{29}$$

and (26) becomes

$$\tilde{\theta}_k = \tilde{\theta}_{k-1} - \frac{1}{c}\,\bar{\phi}_{k-1} D(z)\left[\bar{\phi}_{k-1}^T \tilde{\theta}_k + \theta^T (z^{-1}I - F)^{-1} e_1 \bar{\phi}_{k-1}^T \tilde{\theta}_k\right]$$

or

$$\tilde{\theta}_k = \tilde{\theta}_{k-1} - \frac{1}{c}\,\bar{\phi}_{k-1} H(z)\left[\bar{\phi}_{k-1}^T \tilde{\theta}_k\right], \tag{30}$$

where

$$H(z) = D(z)/A(z). \tag{31}$$

Note the similarities between (30) and (6).

We have the equivalent of Lemma 1 for the output error scheme.

Lemma 2

For the output error algorithm (12), (14), (15) with $\{y_k\}$ and $\{u_k\}$ exactly satisfying (2) and with $H(z)$ strictly positive real (SPR) [i.e., $H(z)$ exponentially stable and $\mathrm{Re}(H(e^{i\omega})) > \delta > 0$ for all real ω], we have the following.

(a) [6]: for $\{\phi_k\}$ bounded,

$$\lim_{k\to\infty}\left[y_k - \bar{\phi}_{k-1}^T \hat{\theta}_{k-1}\right] = 0.$$

(b) [5]: if $\{u_k\}$ is persistently exciting according to (9) and the plant polynomials $A(z)$, $B(z)$ are relatively prime, then $\{\hat{\theta}_k\}$ converges exponentially fast to θ.

(c) [7]: for sufficiently large c the convergence rate of part (b) is linear in c^{-1}, that is, decreasing step size proportionally decreases the convergence rate.

Lemmas 1 and 2 summarize some existing results on the convergence rates for the gradient-based equation error and output error algorithms. Variants of these algorithms can be constructed [1,4,8] that effect a least squares (or Newton-based) adaptive identifier—recursive least squares with exponential forgetting factor is a typical example. The analysis of these schemes is technically more intricate but fundamentally identical to that of the gradient methods. Indeed, the general algorithm structure studied here is common to many other areas of adaptive systems [9], and thus our results are fairly widely applicable. We have concentrated on scalar gain output error methods since these are the least cluttered, meaningful applications of this averaging theory to robustness.

Having presented the algorithm's error equations in "ideal" situations where (2) holds, we shall analyze their nonideal form later in Section V, where we show how, subject to conditions, the departure from nonidealness manifests itself through the inclusion of additive terms driving error equations similar to those above. Robustness is associated with local bounded-input/bounded-output stability of these nonhomogeneous equations, which, in turn, is linked back to exponential stability of the homogeneous equations. Averaging theory will be used to study certain conditions for the existence of these stability properties.

IV. DISCRETE-TIME AVERAGING METHODS

We now distance ourselves temporarily from the immediate questions concerning the stability of adaptive identification to consider general averaging methods for the analysis of slow difference equations.

Consider the linear, time-varying, homogeneous difference equation in $\mathbb{R}^p$

$$x_{k+1} = (I + \epsilon A_k) x_k, \tag{32}$$

where $\{A_k\}$ is a sequence of bounded $p \times p$ matrices and $\epsilon > 0$.

Following the method of [10], we consider the mth iterate of (32):

$$x_{k+m} = \left[\prod_{i=1}^{m} (I + \epsilon A_{k+i-1})\right] x_k. \tag{33}$$

The iterated product of (33) may now be expanded:

$$\prod_{i=1}^{m} (I + \epsilon A_i) = I + \epsilon \sum_{i=1}^{m} A_i + \epsilon^2 \sum_{i=1}^{m} \sum_{j=i+1}^{m} A_j A_i + \cdots$$

$$+ \epsilon^m A_m A_{m-1} \cdots A_1 \tag{34}$$

$$= I + \epsilon m \frac{1}{m} \sum_{i=1}^{m} A_i + (\epsilon m)^2 \frac{1}{m^2} \sum_{i=1}^{m} \sum_{j=i+1}^{m} A_j A_i$$

$$+ \cdots + (\epsilon m)^m \frac{1}{m^m} A_m \cdots A_1. \tag{35}$$

Thus

$$\prod_{i=1}^{m} (I + \epsilon A_{k+i-1}) = I + \epsilon m \frac{1}{m} \sum_{i=1}^{m} A_{i+k-1} + R(k, \epsilon m). \tag{36}$$

Denote by a_∞ the ℓ_∞ norm on $\{A_k\}$; then

$$\|R(k, \epsilon m)\|_\infty \leq \frac{\epsilon^2 m^2 a_\infty^2}{2} \exp(\epsilon m a_\infty). \tag{37}$$

The importance of this result is that, for $\epsilon m a_\infty$ small, we have $R(k, \epsilon m)$ uniformly an order ϵ^2 term in (36) for fixed m. We shall next show that, subject to the summation term in (36) being nondegenerate, the order ϵ term may be chosen to dictate the behavior of (32).

Define the sliding average

$$\bar{A}_k(m) = \frac{1}{m} \sum_{i=1}^{m} A_{k+i-1} \tag{38}$$

and denote by $\mu(A)$ the measure of the matrix A [11]:

$$\mu(A) = \lim_{\beta \to 0} (|I + \beta A|_i - 1)/\beta, \tag{39}$$

where $|\cdot|_i$ is the matrix norm on $\mathbb{C}^{p \times p}$ induced by the vector norm on $\mathbb{C}^p$. For any norm and all A in $\mathbb{C}^{p \times p}$,

$$-\mu(-A) \le \operatorname{Re} \lambda(A) \le \mu(A). \tag{40}$$

Following [10,12], we have the following theorem.

Theorem 1 [10,12]

If there exists an integer m and $\alpha > 0$ such that for all k

$$\mu\left[\bar{A}_k(m)\right] \le -\alpha, \tag{41}$$

or, taking $|x|^2 = x^T P x$ for some fixed, positive-definite matrix P, that for each eigenvalue $\lambda_i(\cdot)$,

$$\lambda_i\left(P\bar{A}_k(m) + \bar{A}_k^T(m) P\right) \le -\alpha, \tag{42}$$

then there exists an $\epsilon_0 > 0$ such that the zero solution of (32) is uniformly (exponentially) asymptotically stable, for all $0 < \epsilon \le \epsilon_0$.

Proof. For uniform asymptotic stability of (32) to hold, we require that the product term in (33) be uniformly contractive in all directions in $\mathbb{R}^p$. That is, we require that for some $\gamma > 0$.

$$\left|\left(\prod_{i=1}^{m} (I + \epsilon A_i)\right) x\right| \le (1 - \gamma)|x| < |x|$$

for any vector x. Adopting the spectral norm, this equates with

$$\left\| \prod_{i=1}^{m} (I + \epsilon A_i) \right\| \le 1 - \gamma,$$

or

$$\|I + (\epsilon m)\bar{A}_k(m)\| \leq 1 - \gamma + \frac{\epsilon^2 m^2 a_\infty^2}{2} \exp(\epsilon m a_\infty).$$

Using (39) and (41) or (42), we may choose γ arbitrarily close to $\epsilon m\alpha$ by allowing ϵm to be sufficiently small. □

Note that, taking $P = \frac{1}{2}I$, condition (42) specializes to a condition on the symmetric part of $\bar{A}_k(m)$.

The correct interpretation of this theorem is that difference equations with sufficiently small updates will have long-term stability properties identical to those of an averaged equation provided this averaged equation exhibits a prescribed (order ϵ) degree of stability/instability. The strong tying together of stability properties follows from the conditions (41) or (42), which force nondegeneracy of the averaged equations's stability. Stability results may also be derived for singular $\bar{A}_k$ using second-order (in ϵ) properties [13]. In continuous time these averaging results are now classical for periodic and almost periodic $\{A(t)\}$ (see, e.g., [14,15]). For processes not possessing true averages, these results recently have been extended in [12]. Equally, the classical averaging theory extends to the analysis of periodic systems that have a range of validity in ϵ that may greatly exceed ours here—Floquet analyses, for example, need not be asymptotic in ϵ. Classical averaging methods for continuous-time systems are very nicely presented by Volosov [15].

We next turn our attention to nonlinear variants of (32). Specifically, we consider

$$x_{k+1} = x_k + \epsilon f(x_k, k) \tag{43}$$

and consider the deviation of $\{x_k\}$ from some nominal trajectory

$\{x_k^*\}$ satisfying the nonlinear homogeneous equation (43)—typically one considers $\{x_k^*\}$ as the zero trajectory and has $f(0, k) = 0$. Using (43) we have

$$x_{k+1} - x_{k+1}^* = x_k - x_k^* + \epsilon f(x_k, k) - \epsilon f\left(x_k^*, k\right), \qquad (44)$$

which we may expand by Taylor's theorem in the neighborhood of x_k^* provided $f(\cdot, k)$ has continuous partial derivatives at x_k^* uniformly in k.

Defining

$$\tilde{x}_k = x_k - x_k^*,$$

(44) becomes

$$\tilde{x}_{k+1} = \tilde{x}_k + \epsilon\left[\nabla_x f\left(x_k^*, k\right)\tilde{x}_k + O|\tilde{x}_k|^2\right], \qquad (45)$$

or

$$\tilde{x}_{k+1} = \tilde{x}_k + \epsilon A_k \tilde{x}_k + \epsilon O|\tilde{x}_k|^2. \qquad (46)$$

We may now use (46) to generalize Theorem 1 to the nonlinear equation (43).

Theorem 2

Consider the nonlinear, time-varying difference equation

$$x_{k+1} = x_k + \epsilon f(x_k, k)$$

with a given solution trajectory $\{x_k^*\}$. If $f(\cdot, k)$ has continuous partial derivatives at x_k^* uniformly in k and we denote the Jacobian matrix

$$(A_k)_{ij} = \left.\frac{\partial f_i}{\partial x^j}\right|_{x_k^*, k},$$

then, if the sliding average of $\{A_k\}$ defined in (38) satisfies (41) for some integer m and $\alpha > 0$, there exists an $\epsilon_0 > 0$ and $\delta > 0$ such that for all $0 < \epsilon \leq \epsilon_0$, $\{x_k\}$ converges exponentially fast to x_k^* provided $|x_0 - x_0^*| < \delta$.

The above theorem represents a linearized averaging principle and describes conditions for combining linearization in a neighborhood of a trajectory with averaging theory to assess the stability of that trajectory (see [16]). There are several points worth noting here. First, as remarked earlier, one typically considers $\{x_k^*\}$ to be the zero trajectory when Theorem 2 represents an exponential asymptotic stability result. Second, from (45) the neighborhood of validity of the theorem depends upon the magnitude of the first partial derivative of $f(\cdot, k)$, but clearly higher order (in $\tilde{x}_k$) stability results could be derived. Third, it is more usual in the classical averaging theory to reverse the above order so that time averaging is performed before the linearization. This occurs primarily because classical results [17] deal chiefly with periodic systems such as answering questions about the stability of limit cycles. Our formulation here is different.

Theorem 2 specifies sufficient conditions for a given solution $\{x_k^*\}$ of (43) to be exponentially stable in the sense of that solution exponentially attracting all other solutions from initial conditions within its own neighborhood. Nothing, however, is stated concerning the asymptotic stability of $\{x_k^*\}$. Clearly, if $\{x_k^*\}$ is an asymptotically stable trajectory, so that $x_k^* \to 0$ as $k \to \infty$, then under the conditions of Theorem 2 $x_k \to 0$ also. In an adaptive system context, it may prove difficult to construct a solution trajectory $\{x_k^*\}$ and even more difficult to ascribe a behavior to it as $k \to \infty$, and so we provide the following result allowing linearization about a fixed point. The importance of Theorem 3 vis-à-vis Theorem 2 is that it does not require the a priori specification of a solution trajectory. Indeed, given the satisfaction of its conditions, it is sufficient for the existence of a trajectory in the neighborhood of the point chosen. Note that there is no presumption of the fixed point being a solution.

Theorem 3

Consider the nonlinear, time-varying difference equation

$$x_{k+1} = x_k + \epsilon f(x_k, k)$$

and consider the point x^*. Suppose $f(\cdot, k)$ has continuous partial derivatives at x^* uniformly in k, and we denote the Jacobian matrix

$$(A_k)_{ij} = \left.\frac{\partial f_i}{\partial x^j}\right|_{x^*,k}.$$

Then, if the sliding average of $\{A_k\}$ defined in (38) satisfies (41) for some integer m and $\alpha > 0$, there exists an $\epsilon_0 > 0$ and $\delta > 0$ such that for all $0 < \epsilon \leq \epsilon_0$, $\{x_k\}$ converges exponentially fast to an $O(\epsilon)$-neighborhood of x^* provided $|x_0 - x^*| < \delta$.

In applying the results of this section to the adaptive identification algorithms of the previous section, we shall equate the parameter ϵ with $1/c$ in (6) and (30) or (25)-(26). Our results then apply to slow adaptation, where the slow variable is the parameter estimate. The averaging techniques applied to these equations should be seen as perturbational, yielding stability results for sufficiently small ϵ and neighborhoods of trajectories, and generally saying nothing whatsoever concerning behavior outside these regions. Nevertheless, the results are not simply asymptotic but reflect a requirement for time-scale separation, which will be familiar to those readers acquainted with a singular perturbation approach to robustness of adaptive systems [18]. This separation in time scale is best seen in the requirements for the averaging result of Theorem 1, where ϵm must be made small. The integer m reflects the coherence length of $\{A_k\}$, so that ϵm being small implies that x_k moves more slowly than A_k.

V. ROBUSTNESS OF ADAPTIVE IDENTIFICATION

In Section II we motivated the study of robustness of adaptive identification and suggested that the qualifier for robust performance was the ability to produce parameter estimates close to a nominal best value in spite of inexact modeling and/or time variation of this nominal value. Our efforts in this section will be first to derive a description of the above deviations from having a single, fixed, exact model (2) as additive perturbations to linearized equations. We will then appeal to the theory of total stability [16,19] to establish that exponential asymptotic stability of the homogeneous parts of these equations plus boundedness of the perturbations implies bounded deviation of the solutions. Robust conditions for exponential stability will then be derived and interpreted in terms of experimental conditions and algorithm gain requirements. Finally, in the next section, modifications to the algorithms are suggested that reinforce their robustness. The key to our development here will be the averaging theory of the previous section. In particular, Theorem 3 will be used to discuss the local behavior of our adaptive systems in the neighborhood of certain fixed parameter values.

Denote by $\left\{\theta_k^*\right\}$ the sequence of parameter values that instantaneously minimizes the model-fitting criterion and denote by $\{e_k\}$ the minimizing error that is no longer assumed to be zero, that is

$$e_k = y_k - \phi_{k-1}^T \theta_k^*. \tag{47}$$

Since θ_k^* need not be constant, we write

$$\delta_k = \theta_{k+1}^* - \theta_k^* \tag{48}$$

and redefine

$$\tilde{\theta}_k = \theta_k^* - \hat{\theta}_k. \tag{49}$$

Our parameter error equations (6) and (30) become

$$\tilde{\theta}_k = \tilde{\theta}_{k-1} - \frac{1}{c}\, \phi_{k-1}\phi_{k-1}^T \tilde{\theta}_k - \frac{1}{c}\, \phi_{k-1} e_k + \delta_{k-1} \tag{50}$$

and

$$\tilde{\theta}_k = \tilde{\theta}_{k-1} - \frac{1}{c}\, \overline{\phi}_{k-1}^* H_k^*(z)\left[\overline{\phi}_{k-1}^* \tilde{\theta}_k\right] - \frac{1}{c}\, \overline{\phi}_{k-1}^* D(z) e_k + \delta_{k-1}, \tag{51}$$

where the slight change (or abuse) in notation is made to reflect the dependence of $\overline{\phi}_{k-1}$ and H(z) on quantities that are now time varying.

Equation (50) is truly linear so that exponential stability of its homogeneous part (6) and boundedness of $\{e_k\}$ and $\{\delta_k\}$ are sufficient to produce bounded $\left\{\tilde{\theta}_k\right\}$. The effect of altering c is also transparent. This has been studied in [20]. The robustness of equation error adaptive identification therefore hinges on the conditions for exponential stability given in Lemma 1, that is, effectively persistence of excitation of the input sequence $\{u_k\}$. The coprimeness of time-varying polynomials $A_k^*(z)$ and $B_k^*(z)$ would be an inappropriate additional condition since this does not necessarily ensure the fundamental underlying requirement for exponential convergence that

$$\rho_1 I \geq \sum_{i=k}^{k+\ell} \phi_i \phi_i^T \geq \rho_2 I > 0, \tag{52}$$

given the persistence of excitation of inputs $\{u_k\}$ alone. In our robustness analysis, however, we shall be able to specify fixed polynomials $A^*(z)$, $B^*(z)$ for which coprimeness will provide this transmission of persistence from $\{u_k\}$ to $\{\phi_k\}$. This condition has also been analyzed in this context by others [21].

To examine robustness of output error methods is more difficult because the dependence of $\left\{\bar{\phi}_k\right\}$ on $\left\{\hat{\theta}_k\right\}$ in both (30) and (51) is nonlinear. For a *fixed* nominal best parameter value θ^*, we may appeal directly to Lemma 2 and Theorem 3 of Section IV to prove that, subject to $\left\{\bar{\phi}_k\right\}$ satisfying a persistence condition like (52) and $H^*(z)$ being SPR, the homogeneous adaptive output error equations are exponentially asymptotically convergent in a neighborhood of θ^* and its associated $\left\{\bar{\phi}^*\right\}$ sequence. Consequently, the output error scheme will be robust to $\{e_k\}$ within certain bounds. Indeed, to establish robustness about a nominal trajectory satisfying the SPR condition on $H^*(z)$, the use of averaging theory is somewhat inappropriate since alternative, more general proofs of exponential convergence exist [22].

The key problem for robustness is, in fact, the satisfaction of the SPR property. Rohrs *et al.* [23] have shown in several examples from an adaptive control context that $H^*(z)$ need not remain SPR for all allowable values of θ_k^*. In particular, it is demonstrated by analysis and simulations that unmodeled high-frequency dynamics in the plant can typically lead to generic inability to satisfy the SPR condition, especially at high frequency. Also, recent work of Wahlberg and Ljung [24] suggests that, with prediction error identification algorithms, the model fitted to the data will actually accentuate high-freqquency fitting to the detriment of low-frequency model matching, thus heightening this problem of non-SPR $H^*(z)$. Here averaging has an important role to play in deriving sufficient conditions for exponential asymptotic stability of the homogeneous error equations for non-SPR $H^*(z)$. The derived conditions are robust in the sense that their satisfaction for a particular stable $H(z)$ guarantees their satisfaction for all neighboring transfer functions. (Neighboring here is in terms of the ℓ_2 norm or its equivalent L_∞ frequency response norm.)

We consider the linearization of (25), (26) about the parameter value θ^* with actual plant regression vectors $\{\phi_k\}$. Then, using the equivalent form (27), (28) and neglecting second-order effects, we have the homogeneous part

$$\tilde{\phi}_k = F(\theta^*)\tilde{\phi}_{k-1} + e_1\phi_{k-1}^T\tilde{\theta}_k, \tag{53}$$

$$\tilde{\theta}_k = \tilde{\theta}_{k-1} - \frac{1}{c}\,\phi_{k-1}D(z)\left[\phi_{k-1}^T\tilde{\theta}_k - \tilde{\phi}_{k-1}^T\theta^*\right], \tag{54}$$

or, using our mixed notation again and denoting $\epsilon = 1/c$,

$$\tilde{\theta}_k = \tilde{\theta}_{k-1} - \epsilon\phi_{k-1}H(z)\left\{\phi_{k-1}^T\tilde{\theta}_k\right\} \tag{55}$$

Theorem 4

Consider (55) and suppose that the transfer function $H(z)$ is exponentially stable and that the sequence of vectors $\{\phi_k\}$ is bounded; then there exists an ϵ_0 such that for all $0 < \epsilon \leq \epsilon_0$ (55) is exponentially asymptotically stable if

$$\tilde{\theta}_k = \tilde{\theta}_{k-1} - \epsilon\phi_{k-1}H(z)\left\{\phi_{k-1}^T\right\}\tilde{\theta}_{k-1} \tag{56}$$

is exponentially asymptotically stable.

N.B. Equation (56) differs from the exact error system (55) in two crucial ways. First, it is now a strictly causal difference equation with $\tilde{\theta}_{k-1}$ only appearing on the right-hand side. Second, the operator in (55) has been replaced by a memoryless time-varying operator in (56). Equation (56) therefore has the form of (32), to which averaging may be applied.

However, in (56) there is still the nonlinear connection between $\{\theta_k\}$ and $\{\phi_k\}$ since ϕ_k contains $\hat{y}_k$. To complete the setting up for averaging analysis, we shall later linearize (56) to replace $\hat{y}_k$ by y_k.

Proof. Denote the impulse response of H(z) by $\{h_0, h_1, h_2, \ldots\}$, then

$$\begin{aligned} H(z)\left\{\phi_{k-1}^T\tilde{\theta}_k\right\} &= h_0\phi_{k-1}^T\tilde{\theta}_k + h_1\phi_{k-2}^T\tilde{\theta}_{k-1} + h_2\phi_{k-3}^T\tilde{\theta}_{k-2} + \cdots \\ &= \left\{h_0\phi_{k-1}^T\tilde{\theta}_{k-1} + h_1\phi_{k-2}^T\tilde{\theta}_{k-1} + h_2\phi_{k-3}^T\tilde{\theta}_{k-1} + \cdots\right\} \\ &\quad + \left\{h_0\phi_{k-1}^T\left(\tilde{\theta}_k - \tilde{\theta}_{k-1}\right) + h_2\phi_{k-3}^T\left(\tilde{\theta}_{k-2} - \tilde{\theta}_{k-1}\right) \right. \\ &\quad \left. + \cdots\right\}. \end{aligned} \tag{57}$$

We now bound the second term above as an order of ϵ. Denote by γ the overbound on $|\phi_k|$, and by m the ℓ_∞ gain of H(z); then from (55) we have

$$\sup_{i<k}|\tilde{\theta}_i| \le (1 + \epsilon\gamma^2 m)^k K_0, \tag{58}$$

where K_0 represents the initial condition magnitude on $\tilde{\theta}$ and the state of H(z). Further, using (55) and then (58), we have

$$|\tilde{\theta}_k - \tilde{\theta}_{k-n}| \le \epsilon n\gamma^2 m \sup_{i<k}|\tilde{\theta}_i| \le \epsilon n\gamma^2 m(1 + \epsilon\gamma^2 m)^k K_0.$$

Since the impulse response sequence $\{h_i\}$ is exponentially decaying to zero, we may choose ϵ sufficiently small for the second term in (57) to be bounded as

$$\left|H(z)\left\{\phi_{k-1}^T\tilde{\theta}_k\right\} - H(z)\left\{\phi_{k-1}^T\right\}\tilde{\theta}_{k-1}\right| \le O(\epsilon) \sup_{i<k}|\tilde{\theta}_i|. \tag{59}$$

The result follows using the device of the Appendix of [25] or appealing to the Small Gain Theorem [11]. □

The impact of Theorem 4 is that it converts our analysis of the linearized homogeneous error equations of output error adaptive identification from the form of (55) or its state-variable equivalent to an analysis directly of linear time-varying equations in the form amenable to averaging methods studied in Section IV. The study of state-variable equations proceeds formally along lines similar to those above in that one attempts to separate the fast dynamics of H(z) from the slow dynamics of $\tilde{\theta}$. These slow dynamics may then be analyzed by averaging. This has been the method of attack utilized by Riedle and Kokotovic [26,27], where separation of time scales admits the isolation of the dynamics of individual modes using Lyapunov transformations reminiscent of singular perturbation theory [28].

The modified equation (56) is now of the form

$$\tilde{\theta}_k = \tilde{\theta}_{k-1} - \epsilon A_k \tilde{\theta}_{k-1},$$

where

$$A_k = \phi_{k-1} H(z)\left\{\phi_{k-1}^T\right\}. \tag{60}$$

According to our earlier analysis of Theorem 1, we require that, for bounded $\{A_k\}$ (which we have here by assumption), the sliding average values $\bar{A}_k(m)$ defined by (38) should satisfy a condition like (41) or (42) and that ϵ should be small. Specifically, we have the following theorem.

Theorem 5

There exists an $\epsilon_0 > 0$ such that for all $0 < \epsilon \leq \epsilon_0$ the homogeneous, linearized output error adaptive identification error equation (55) is exponentially asymptotically stable if $\{\phi_k\}$ satisfies the condition that for some integer m, positive constant α, and all k,

$$\lambda_i\left[\bar{A}_k(m) + \bar{A}_k^T(m)\right] \geq \alpha > 0, \tag{61}$$

where

$$\overline{A}_k(m) = \frac{1}{m} \sum_{j=1}^{m} \phi_{k+j} H(z) \left\{ \phi_{k+j}^T \right\}$$

and where H(z) is evaluated at θ^*.

The condition (61) from Theorem 5 represents an average positivity condition on $\{\phi_k\}$ and H(z). It in general implies restrictions on both $\{\phi_k\}$ and on H(z). If H(z) is strictly positive real, then the passivity of H(z) implies (assuming zero initial conditions) that the sum in braces in (61) is nonnegative for all k and m. Thus, (61) is a stronger condition since the sum must be uniformly strictly positive definite. To achieve this property with SPR H(z) requires $\{\phi_k\}$ to be persistently exciting (see [5,20]), that is, $\{\phi_k\}$ must satisfy (53). Since $\{\phi_k\}$ is a vector sequence composed of inputs and outputs to the actual plant, the persistence of excitation requirement may be devolved to a persistence condition on the input process alone plus a coprimeness constraint on the plant [5,29]. Further, this coprimeness restriction can be transferred to the nominal plant provided we consider parameter values in a neighborhood of the nominal value that preserves coprimeness—this is easily accommodated [30]. We should note here that with the transposition of conditions on $\{\phi_k\}$ to conditions on the inputs and outputs of the actual plant, we have not only avoided the nonlinear problems but, at the same time, sidestepped the boundedness question. In other words, if the nominal plant is exponentially stable and the input is bounded, then, in our analysis, subject to other conditions, $\{\phi_k\}$ will be bounded.

When H(z) is not SPR it is still possible to achieve satisfaction of (61), but subject to correct choice of $\{\phi_k\}$ and subject still to persistence of excitation. Following Riedle and Kokotovic [26], we may consider the particular instance of m-periodic $\{\phi_k\}$, or

$$\phi_k = \sum_{i=0}^{m-1} \beta_i \exp(j^{2\pi ik}/m),$$

to rewrite condition (61) as

$$\sum_{i=0}^{m-1} \beta_i H[\exp(j^{2\pi i}/m)]\beta_i^* > \alpha I, \tag{62}$$

which demonstrates that the stability requirement is not that $\text{Re}\{H[\exp(j\omega)]\} > 0$ for all ω but rather that $\{\phi_k\}$ have its significant frequency content where the frequency response of H(z) has a positive real part. This conclusion has been derived using small-gain analysis in [31] to produce possibly nonasymptotic (in ϵ) stability results for equations such as (55). Their analysis is not strictly averaging theoretic, however.

We are now in a position to return specifically to the robustness of adaptive identification issue, where e_k—the minimal modeling error (47)—need not be zero but enters our analysis as an additive driving term to our homogeneous error equations. Using the theory of total stability [16,19], we have the following theorem.

Theorem 6

Consider the output error adaptive identification algorithm (12), (14), (15) and consider a nominal parameter estimate θ^* defining the transfer function $H^*(z)$ via (31). Let $\{\phi_k\}$ be the sequence of measured information vectors (3) of the actual plant with inputs $\{u_k\}$ and outputs $\{y_k\}$ and define

$$e_k = y_k - \phi_{k-1}^T\theta^*. \tag{63}$$

If

(1) $H^*(z)$ is exponentially asymptotically stable and has no pole-zero cancellations;

(2) $\{u_k\}$ is bounded and is persistently exciting according to (9);

(3) $\{\phi_k\}$ and $H^*(z)$ satisfy the average positivity condition (61) for some $\alpha > 0$ and all k;

then there exist $\epsilon_0 > 0$, $\delta > 0$, $E > 0$ such that

(a) $0 < \epsilon \leq \epsilon_0$;

(b) $0 < |\theta_0 - \theta^*| \leq \delta$;

(c) $0 < |e_k| \leq E$;

imply $\hat{\theta}_k - \theta^*$ and $\overline{\phi}_k - \phi_k$ are bounded for all k.

There is a direct extension of these results to the case of time-varying nominal values $\left\{\theta_k^*\right\}$ using the results of [21].

Corollary 1

If we consider a possibly time-varying sequence of nominal parameter values $\left\{\theta_k^*\right\}$ on a compact set and a corresponding sequence of transfer functions $\left\{H_k^*(z)\right\}$ uniformly satisfying conditions (1), (2), and (3) of Theorem 2, then defining

$$e_k = u_k - \phi_{k-1}^T\theta_k^*,$$

there exist ϵ_0, δ, E, and Δ such that (a), (b), (c), and (d), $|\theta_k^* - \theta_{k-1}^*| \leq \Delta$ for all k, imply $\hat{\theta}_k - \theta_k^*$ and $\overline{\phi}_k = \phi_k$ are bounded for all k.

Theorem 6 and Corollary 1 encapsulate the main robustness results that we have derived here via the averaging theory of Section IV. In the next section we move on to discuss the design and operational implications of these results for achieving robust adaptive identification.

VI. IMPLICATIONS FOR DESIGN AND OPERATION

Tracing our path through to this point of the article, we have developed technical results that guarantee local bounded-input/bounded-output (BIBO) stability of the equations describing the parameter and state error of adaptive identification algorithms. As was motivated in Section V and developed in [32,33], the plant uncertainty assumed to underlie the adaptive identification problem enters the local analysis as an additive driving term to the ideal equations, permitting exact model-plant matching. Robustness is equated with this local BIBO stability, and so our questions now turn to considering the implications of the conditions of the theorems of Section V in terms of system or algorithm design properties. Historically, the examples of Rohrs *et al.* [23] demonstrating nonrobustness in adaptive control spurred a considerable activity in the study of the mechanisms of robustness. Aström [34] gave a penetrating analysis of one particular example and showed by heuristics, supported by simulation, how a level of robustness can be achieved. His results concur strongly with some of our own to be presented here, and we cite [34] as a valuable connection between this work and the current robustness debate.

A. SMALL STEP SIZE ϵ

As has been frequently stated, the nature of the theory of averaging applied here dictates that ϵ, the algorithm step size, be small and this, in turn, must be perceived as requiring that the time constant of the adaptation must be significantly longer than that of the plant. That is not to say that adaptive identification with large step size will not be robustly stable, but it does suggest very strongly, by its very averaging nature, that any analysis of nonlinear adaptive processes on the basis of linear certainty-equivalence-styled assumptions should be predicated on slow adaptation. Within the framework of Section II, where we postulate adaptive estimation as a variant of system identification designed to accommodate plant uncertainty that is possibly time varying, this time scale separation is tied to the validity of the ARMA models being proposed and to the use of modified, previously asymptotically consistent parameter estimators in adaptive identification. Having ϵ small allows the estimator to average over the plant response to identify the parameter.

B. *PERSISTENCE OF EXCITATION*

Persistence of excitation of signals is intimately connected to questions of identifiability and for standard identification algorithms is a requirement (in a slightly modified form) for consistency [35,36]. In the realm of adaptive identification, where one uses parameter estimators that effectively locally window the data, persistence of excitation corresponds to uniformity of identifiability of the parameters and so is an unavoidable requirement. Without persistence of excitation, under certain circumstances it is possible to demonstrate complete instability of even simple adaptive schemes [37,38]. The coprimeness of models arises only as a concern to ensure that persistence of the input sequence $\{u_k\}$ is sufficient to force persistence of the complete regression vector.

There have been several algorithms suggested for situations where persistence does not hold. These algorithms typically rely on projection, normalization, etc., and are no longer members of the class of algorithms being considered here. For algorithms of our class, we believe persistence of excitation to be practically a necessary condition for robustness.

C. NOMINAL MODEL STABILITY

Insisting that H(z), for the nominal parameter value, be asymptotically stable is not unreasonably restrictive, since the intention of adaptive identification in this context with small ϵ is to adjust incrementally the parameter estimate to track closely a slowly time-varying nominal best value. Stability of H(z) corresponds to stability of the nominal plant since both have denominator polynomial A(z). Identification of slowly time-varying unstable plants is not in the province of this article and is technically somewhat more difficult.

D. AVERAGE POSITIVITY AND THE SPR CONDITION

The average positivity requirement (61) has also been discussed earlier and represents a condition that replaces the SPR property by a much less restrictive signal-dependent one. The signal dependence here admits a much broader class of H(z) than SPR for given signal $\{\phi_k\}$, although persistence of excitation remains a necessity. There are clearly two ways to attempt to alter the adaptive identification method to try to encourage the satisfaction of (61)—changing $\{\phi_k\}$ and changing the nominal parameter value and hence H(z).

The particular sequence of regressor vectors $\{\phi_k\}$ is clearly a direct function of the input signal $\{u_k\}$, and thus this aspect of the average positivity condition is available for selection and intelligent choice. If one knows the approximate frequency response of $H(z)$, then the input may sometimes be able to be tailored to maintain persistence of excitation and to satisfy (61). This has been analyzed in some detail in [26,31].

In case the frequency response of $H(z)$ is not well known, we still have two design choices. First, the error-filtering polynomial $D(z)$ is a design variable with which it is possible to encourage, say, low-pass $H(z)$ by placing the zeros. Second, we may alter the nominal parameter value to one that better suits the achievement of average positivity. The available methods to alter optimizing parameter values are essentially to change either the experimental conditions or the estimation criterion [2]. The former is already somewhat constrained by persistence requirements but admits some latitude for change. Changing the estimation criterion is also interesting since it implies a modification of the explicit algorithm. Using a standard prediction error estimation criterion may lead to an emphasis on high-frequency model fitting [24], which, typically, is entirely inappropriate for control systems identification. Wahlberg and Ljung [24] propose using prefilters on the signals of the algorithm to enhance fitting in a more desirable frequency band. This same approach is used in an adaptive control context in [39,40], specifically to force approximate satisfaction of the SPR condition at low frequencies, which is the target range for practical control improvement.

E. CONCLUSIONS

We have now presented an application of the theory of averaging to the analysis of the robustness of adaptive identification. Our particular focus has been the output error scheme since this is representative of a large class of algorithms, demonstrating the nonlinear and time-varying properties but also being relatively simple in structure. (Extensions to adaptive control and more complicated algorithms such as variants of recursive least squares are possible.) The outcome of this study has been to derive conditions to guarantee the performance of slow adaptive identification and, in turn, these conditions have been interpreted in terms of their providing guidelines to actions likely to enhance robustness of adaptive identification.

The chief principles for this theory to predict robustness enhancement are small step size/slow adaptation, persistence of excitation, suitable error filter D(z) choice, introduction of prefilters on the signals to the identifier to encourage the average positivity, and concurrent choice of frequency content of the plant input signal. In this way, the results of this article attempt to indicate suitable rules for modification of essentially all the free variables available to the designer/operator of the adaptive identifier.

ACKNOWLEDGMENTS

The technical advice and comments of Iven Mareels were of great benefit in formulating and prosecuting this article, while "l'échange libre" established with our other colleagues and friends (in no particular order), Brian Anderson, Brad Riedle, Petar Kokotovic, Robert Kosut, Laurent Praly, and Bill Sethares, has proven invaluable in the generation of ideas and the development of theory and technique. To add a note of historical perspective, the germs of many of the ideas intoned here were developed at the Montana Summer Institute on Adaptive Control organized by Don Pierre and attended by the authors and most of the protagonists listed above. Finally, we thank Marie-Antoinette Poubelle for dealing with the technical illustrations.

This work was supported by the US/Australia Cooperative Science Program and National Science Foundation Grant ECS-8119312.

REFERENCES

1. G. C. GOODWIN and R. L. PAYNE, "Dynamic System Identification: Experiment Design and Data Analysis," Academic Press, New York, 1977.

2. L. LJUNG and T. SÖDERSTRÖM, "Theory and Practice of Recursive Identification," MIT Press, Cambridge, Massachusetts, 1983.

3. B. D. O. ANDERSON and R. M. JOHNSTONE, *Proc. Conf. Control Eng., 2nd, Newcastle, New South Wales, Australia*, 59 (1982).

4. G. C. GOODWIN and K. S. SIN, "Adaptive Filtering, Prediction and Control," Prentice-Hall, Englewood Cliffs, New Jersey, 1984.

5. B. D. O. ANDERSON and C. R. JOHNSON, JR., *Automatica 18*, 1 (1982).

6. I. D. LANDAU, *IEEE Trans. Autom. Control AC-21*, 194 (1976).

7. B. D. O. ANDERSON, *Syst. Control Lett. 4*, 119 (1984).

8. D. A. LAWRENCE, C. R. JOHNSON, JR., and B. D. O. ANDERSON, *Proc. Conf. Inf. Syst. Sci.* (1983).

9. C. R. JOHNSON, JR., *IEEE Trans. Autom. Control AC-25*, 697 (1980).

10. R. R. BITMEAD, B. D. O. ANDERSON, and T. S. NG, *Proc. IFAC Triennial World Congr., 9th, Budapest*, 60 (1984).

11. C. A. DESOER and M. VIDYASAGAR, "Feedback Systems: Input-Output Properties," Academic Press, New York, 1975.

12. R. L. KOSUT, B. D. O. ANDERSON, and I. M. Y. MAREELS, *IEEE Trans. Autom. Control AC-32* (1987), in press.

13. C. ROBINSON, *IEEE Trans. Circuits Syst. CAS-30*, 591 (1983).

14. J. K. HALE, "Ordinary Differential Equations," Krieger, Molaban, Florida, 1980.

15. V. M. VOLOSOV, *Russ. Math. Surv. 17*, 1 (1962).

16. T. YOSHIZAWA, "Stability Theory and the Existence of Periodic Solutions and Almost Periodic Solutions," Springer-Verlag, Berlin and New York, 1975.

17. N. N. BOGOLIUBOV and Y. A. MITROPOLSKY, "Asymptotic Methods in the Theory of Nonlinear Oscillations," Gordon & Breach, New York, 1961.

18. P. IOANNOU and P. V. KOKOTOVIC, "Adaptive Systems with Reduced Models," Springer-Verlag, Berlin and New York, 1983.

19. W. HAHN, "Stability of Motion," Springer-Verlag, Berlin and New York, 1967.

20. R. R. BITMEAD and B. D. O. ANDERSON, *IEEE Trans. Autom. Control AC-25*, 788 (1980).

21. B. D. O. ANDERSON and R. M. JOHNSTONE, *Int. J. Control 37*, 367 (1983).

22. B. D. O. ANDERSON and R. M. JOHNSTONE, *Proc. 20th IEEE Conf. Decision Control, San Diego, California*, 510 (1981).

23. C. E. ROHRS, L. VALAVANI, M. ATHANS, and G. STEIN, *Proc. 20th IEEE Conf. Decision Control, San Diego, California*, 1272 (1981).

24. B. WAHLBERG and L. LJUNG, *Proc. 23rd IEEE Conf. Decision Control, Las Vegas, Nevada*, 335 (1984).

25. K. S. NARENDRA and Y.-H. LIN, *IEEE Trans. Autom. Control AC-25*, 456 (1980).

26. B. D. RIEDLE and P. V. KOKOTOVIC, *Proc. 23rd IEEE Conf. Decision Control, Las Vegas, Nevada*, 998 (1984).

27. B. D. RIEDLE and P. V. KOKOTOVIC, *IEEE Trans. Autom. Control AC-31*, 316 (1986).

28. P. V. KOKOTOVIC, *SIAM Rev. 26*, 501 (1984).

29. S. BOYD and S. S. SASTRY, *Proc. Am. Control Conf., 1984, San Diego, California*, 1584 (1984).

30. R. W. BROCKETT, *IEEE Trans. Autom. Control AC-21*, 449 (1976).

31. B. D. O. ANDERSON, R. R. BITMEAD, C. R. JOHNSON, JR., and R. L. KOSUT, *Proc. 23rd IEEE Conf. Decision Control, Las Vegas, Nevada*, 1286 (1984).

32. R. L. KOSUT, *Proc. IFAC Workshop, San Francisco, California*, June 1983.

33. R. L. KOSUT and C. R. JOHNSON, JR., *Automatica 20*, 569 (1984).

34. K. J. ÅSTRÖM, *Proc. 22nd IEEE Conf. Decision Control, San Antonio, Texas*, 982 (1983).

35. V. SOLO, *IEEE Trans. Autom. Control AC-24*, 958 (1979).

36. J. B. MOORE, *IEEE Trans. Autom. Control AC-28*, 60 (1983).

37. W. A. SETHARES, D. A. Lawrence, C. R. JOHNSON, JR., and R. R. BITMEAD, *Proc. Conf. Inf. Syst. Sci., 19th, Baltimore, Maryland* (1985).

38. K. J. ASTRÖM, *Proc. 23rd IEEE Conf. Decision Control, Las Vegas, Nevada*, 1276 (1984).

39. C. R. JOHNSON, JR., B. D. O. ANDERSON, and R. R. BITMEAD, *Proc. 23rd IEEE Conf. Decision Control, Las Vegas, Nevada*, 993 (1984).

40. B. D. O. ANDERSON, R. R. BITMEAD, C. R. JOHNSON, JR., and R. L. KOSUT, *Proc. IFAC Symp. Identif. Syst. Parameter Estim., 7th, York, England*, 1003 (1985).

Techniques for Adaptive State Estimation through the Utilization of Robust Smoothing

F. DALE GROUTAGE

Naval Ocean Systems Center
San Diego, California 92152

RAYMOND G. JACQUOT
R. LYNN KIRLIN

Department of Electrical Engineering
University of Wyoming
Laramie, Wyoming 82070

I. INTRODUCTION

Estimation of the state variables of a dynamic system based on observed data has been a focal point in recent system theoretic literature. The concepts of least squares parameter evaluation and curve fitting were introduced by Gauss in the early nineteenth century [1]. More recently, Norbert Wiener was asked to solve the problem of formulation and specification of optimal properties of servomechanisms [2]. Wiener gave a general solution to this problem based on rigorous probabilistic approaches, and today the Wiener filter is acclaimed as the cornerstone of modern estimation theory. Over the past two decades, modern sequential estimation theory has matured, after Kalman [3] and Kalman and Bucy [4] presented a new, more general formulation of the estimation problem. The language of this theory was that of the so-called modern control theory.

Engineers and scientists given the task of estimating the behavior of a physical system have found themselves relying heavily on modern estimation theory. The general problem is not trivial, since the choice of the estimation technique is often problem dependent. There are many criteria for specification of the methodology for estimating a parameter or variable, with three possible choices being (a) the most probable estimate, (b) the conditional mean estimate, and (c) the minimax estimate [5].

Many derivations of the conventional discrete-time Kalman filter are given in the literature, and one of these is given in [6]. The system of interest is governed by the stochastic matrix-vector difference equation

$$x(k + 1) = Ax(k) + w(k) + Bu(k), \tag{1}$$

with a linear measurement sequence defined by

$$z(k) = Hx(k) + v(k), \tag{2}$$

where $x(k)$ and $w(k)$ are n vectors and $u(k)$ and $z(k)$ are, respectively, j and r vectors. All matrices are of the correct dimensions to have the products and sums of (1) and (2) defined. In this conventional state estimation problem, $w(k)$ and $r(k)$ are independent Gaussian random sequences with respective means $q(k)$ and $r(k)$ and covariances $Q(k)$ and $R(k)$.

The problem is that of finding the optimal estimate of $x(k)$ [called $\hat{x}(k)$] based on the measurement set $Z^k = \{z(1), z(2), \ldots, z(k)\}$ that minimizes the conditional mean square error J given by

$$J = \frac{1}{2} E\{[x(k) - \hat{x}(k)^T][x(k) - \hat{x}(k)] | Z^k\}. \tag{3}$$

Minimization of J gives an estimator of the form

$$\overline{x}(k) = A\hat{x}(k - 1) + q(k - 1) + Bu(k - 1), \tag{4}$$

in which $\overline{x}(k)$ is the extrapolated estimate of $x(k)$ and $\hat{x}(k)$ is the updated estimate of $x(k)$ based on the measurement $z(k)$ given by

$$\hat{x}(k) = \overline{x}(k) + K(k)[z(k) - HA\hat{x}(k-1) - r(k) \\ -HBu(k-1) - Hq(k-1)], \tag{5}$$

where the gain matrix $K(k)$ is

$$K(k) = \overline{P}(k)H^T[H\overline{P}(k)H^T + R(k)]^{-1} \tag{6}$$

and

$$\overline{P}(k) = A\hat{P}(k-1)A^T + Q(k) \tag{7}$$

with

$$\hat{P}(k) = \overline{P}(k) - K(k)H\overline{P}(k). \tag{8}$$

The observation residual is given by

$$y(k) = z(k) - H\overline{x}(k). \tag{9}$$

$R(k)$ is the covariance of the measurement sequence $v(k)$ defined by the expression

$$R(k) = E[(v(k) - r(k))(v(k) - r(k))^T], \tag{10}$$

and $r(k)$ is the mean value of $v(k)$. In a similar fashion, $Q(k)$ and $q(k)$ are, respectively, the covariance and mean of the process noise sequence $w(k)$.

The conventional Kalman filter performs well when operating within the assumptions made in its derivation; however, when the system is subjected to unknown, deterministic forcing functions, the state estimates will diverge from the true state. This is particularly the case when estimating the location of an aircraft or missile undergoing an evasive maneuver.

Figure 1 presents a block diagram of a system model, measurement system, and discrete Kalman filter. The system model could be a discrete representation of a continuous system being

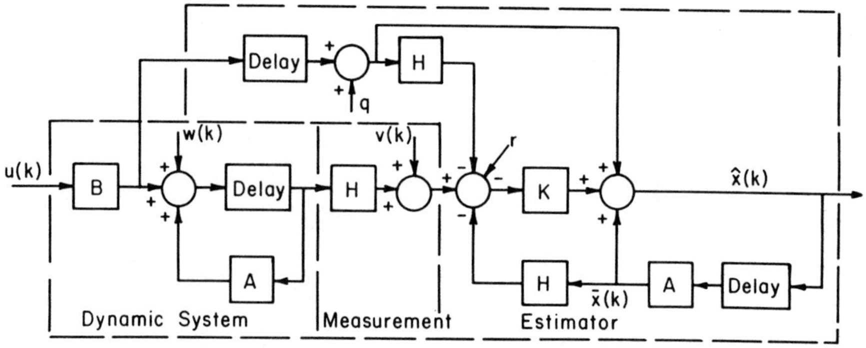

Fig. 1. System model and conventional Kalman filter. (Reproduced from [10] with permission of the ASME.)

observed at discrete times by a measurement system. The assumptions in this case are that u(k) and w(k) are constant over the sampling interval, that is,

$$w(t) = w(k), \quad kT \leqq t \leqq (k + 1)T, \tag{11}$$

and

$$u(t) = u(k), \quad kT \leqq t \leqq (k + 1)T, \tag{12}$$

where T is the sampling interval. A further assumption is that knowledge of the forcing function u(k) exists; thus the Kalman filter, as illustrated in Fig. 1, reflects this knowledge. Consider now the case where a system is being forced by forcing functions u(k) and w(k) and the output is corrupted by noise to produce measurement z(k). The deterministic forcing function u(k) often changes rapidly with time. It is desired to formulate estimates of the value of the state variables in a timely manner by using a Kalman filter; however, for this case, knowledge of the system forcing function u(k) is unknown to the state estimator. A problem of equal importance is the case where the statistics of the process noise w(k) are unknown. An excellent treatment of estimation in the presence of unknown noise statistics is presented in Myers and Tapley [7]. Empirical estimators are developed in [7] that estimate the noise statistics, and these are presented in Appendix A. This appendix presents not only the empirical estimators of Myers and Tapley, which assume Gaussian distributed populations, but also robust empirical estimators of noise statistics that are not constrained to the Gaussian assumption. Simulations have verified that the robust estimators of Appendix A reduce the level of mean-square error in the estimates of noise statistics by as much as 100 times when compared to the same estimates using a conventional estimator under the Gaussian assumption. It is this evidence that has spurred the interest in pursuing the development of robust estimators of statistics.

Fitzgerald [8] discusses divergence of the Kalman filter where insufficient care in modeling the system leads to unacceptable results. Modeling errors are unaccounted for in the estimation algorithm that leads to filter divergence. The explanation for this phenomenon is that the calculated covariance matrix becomes unrealistically small, and thus the filter gain becomes small, so that undue confidence is placed in the estimates and subsequent measurements are ignored. Groutage [9,10] examines the tracking problem and filter divergence resulting from modeling errors in observation and process noises and deterministic system forcing functions. Modeling errors of the associated noises (observation and process) lead to unrealistically small estimate covariance matrices. Errors in modeling of deterministic forcing functions, similarly, lead to filter divergence.

Singer [11] developed a target model in which target maneuvers are modeled as random events. Turns, evasive maneuvers, and accelerations due to atmospheric turbulences are viewed as perturbations upon the constant velocity trajectory and are commonly modeled as process noise. Recently Berg [12] addressed the maneuvering target problem where the prediction of future target position, in terms of estimated target-related parameters such as thrust, lift, and roll rate, is estimated through the solution of a set of nonlinear differential equations. Berg follows Singer's approach, where target acceleration is modeled as a zero-mean, time-correlated random process; however, an additional term is added to the Singer model to enhance maneuvering target estimator performance.

If a target maneuver is viewed as being deterministic rather than as a stochastic process, the problem is then one of determining the point at which the maneuver commences and the nature and effects of the maneuver. For instance, if the acceleration can be interpreted as a step function, the problem is one of step detection. Similarly, if the acceleration is sinusoidal, the problem is sinusoidal detection. Furthermore, if the acceleration and associated maneuver are finite, modeling the maneuver as process noise will keep the filter bandwidth open during the nonmaneuver time periods, with a resultant increase of mean-square error in the estimates of target location.

Adaptive filtering procedures can be used to address the problem of opening and closing the filter bandwidth to accept or reject observation or sensor data. Mehra [13] presents a procedure for estimating the covariance of both the observation and process noises. The concept is based on Kailath's work [14], which shows that a necessary and sufficient condition for optimality of a Kalman filter is that the innovation sequence be white. Mehra's approach (a) checks the Kalman filter constructed using some estimates of the noise covariances to determine how close the filter is to optimum, (b) obtains unbiased estimates of the noise covariances, and (c) adapts the Kalman filter at regular intervals using all previous information. Involved in Mehra's approach is the solution of a set of equations not linearly independent. Like Mehra, Jazwinski [15] presents an adaptive procedure for estimating the covariance matrix of the process noise. Jazwinski's approach is based on the statistics of predictive residuals. The filter is adaptive in the sense that as long as the residuals remain within their $\sigma = 1$ limit, the measurement noise covariance is treated as zero. This is because the residuals are small and consistent with their statistics. When the residuals become large relative to their $\sigma = 1$ values, the filter gains are increased, and the filter is "open" to incoming observations. Two limitations of the adaptive filter are pointed out in Jazwinski [15], namely, that estimates have a positive bias and estimates will respond to measurement noise. It is noted in Olmstead and Goheen [16] that residuals can be large when a target maneuvers. Unfortunately, residuals can also be large for large but natural fluctuations in the measurements. Thus there is a trade-off between (a) making the algorithm sensitive to target maneuvers and letting it be influenced by noisy sensor data and (b) decreasing the maneuver sensitivity so that noisy data can be rejected.

This treatment of the problem, incomplete knowledge of the system forcing function, presents a combination of some of the ideas described in Myers and Tapley [7], Mehra [13], Jazwinski [15], and Olmstead and Goheen [16] in the formulation of an adaptive procedure that not only responds to target maneuvers but also treats noisy data in a meaningful way. The procedure is simple and robust in that it responds to a variety of underlying system inputs (both deterministic and nondeterministic).

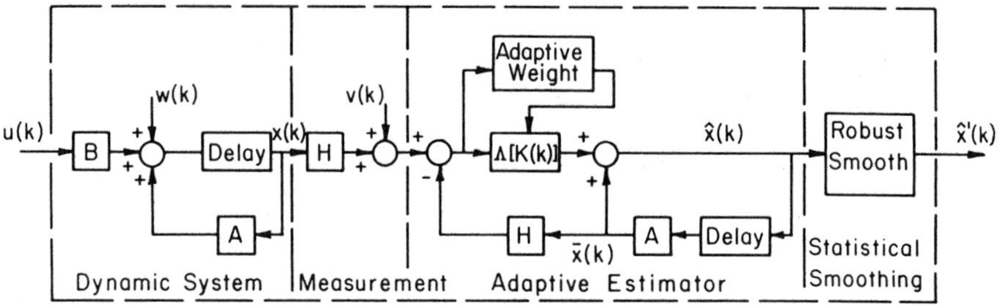

Fig. 2. Modified estimator for adaptive robust estimation. (Reproduced from [10] with permission of the ASME.)

Basically it includes adaptive weighting of the elements of the conventional Kalman gain and covariance matrices, consistent with the residual statistics, as well as robust statistical smoothing of the estimates made by the adaptive Kalman filter using the modified gain and covariance matrices. Adaptive weighting of the gain and covariance matrices regulates the filter's bandwidth to track maneuvers, whereas robust smoothing addresses reduction of mean-square error in the estimates caused by large fluctuations in the measurements. The concept is demonstrated in terms of a step detection example. Details of the robust statistical procedure are presented in Bickel [17] and Huber [18]. The scheme is illustrated conceptually in Fig. 2, where the adaptive weighting and robust statistical smoothing are clearly shown.

II. ROBUST ESTIMATION OF OBSERVED STATE VARIABLES

The concepts presented here are empirical in nature since they are based on observations and experimental data. An intuitive perception of these concepts can be obtained by examining the Kalman filtering algorithm for a scalar system with process noise w(k),

$$x(k+1) = ax(k) + w(k), \tag{13}$$

and the associated noisy measurements,

$$z(k) = x(k) + v(k). \tag{14}$$

Both w(k) and v(k) are assumed to be stationary Gaussian random sequences with variances Q and R, respectively. The Kalman filter equations for estimation of x(k) are, respectively, the propagation equations

$$\bar{x}(k) = a\hat{x}(k-1), \tag{15}$$

$$\bar{P}(k) = a^2\hat{P}(k-1) + Q, \tag{16}$$

the gain equation

$$K(k) = \bar{P}(k)/[\bar{P}(k) + R], \tag{17}$$

and the estimation equations

$$\hat{x}(k) = \bar{x}(k) + K(k)[z(k) - \bar{x}(k)], \tag{18}$$

$$\hat{P}(k) = [1 - K(k)]\bar{P}(k). \tag{19}$$

Equation (18) for estimating the state variable x(k) contains interesting information concerning the estimation process. Note that the scalar gain K(k) is bounded from above and below as

$$0 \leq K(k) \leq 1. \tag{20}$$

For the case in which K(k) = 0, (18) indicates that total faith is placed in the estimation process. In fact, the measurements are ignored and the previous estimate is the updated estimate. Now consider the case in which K(k) = 1, which indicates that there will be no faith in the estimation process and, in fact, the current measurement is the updated estimate. With these concepts in mind, the idea of a pseudo-gain $\alpha(k)$ is investigated, where

$$\alpha(k) = 1 + e^{-\beta(k)}K(k) - e^{-\nu\beta(k)}, \tag{21}$$

and where ν and $\beta(k)$ are parameters to be established and ν can be a constant. The observation residual $y(k)$ is defined as the difference between the measurement and the propagated state estimate:

$$y(k) = z(k) - \bar{x}(k). \tag{22}$$

Now consider the recursive sample mean of the observation residual sequence and the recursive sample variance of this residual sequence. Let the averaging interval be of length N_ℓ and let $\bar{y}(k)$ designate the recursive sample mean of the observation residual sequence so that

$$\bar{y}(k) = \frac{1}{N_\ell} \sum_{j=k-N_\ell+1}^{k} y(j). \tag{23}$$

The recursive sample variance of this residual is derived in Appendix B and is

$$\hat{\sigma}_y^2(k) = \hat{\sigma}_y^2(k-1) + \frac{1}{N_\ell - 1}\Big\{[y(k) - \bar{y}(k)]^2 - \big[y(k - N_\ell) - \bar{y}(k)\big]^2 + \frac{1}{N_\ell}[y(k) - y(k - N_\ell)]^2\Big\}. \tag{24}$$

If the parameter $\beta(k)$ of (21) is chosen as

$$\beta(k) = \gamma\hat{\sigma}_y^2(k), \tag{25}$$

then $\alpha(k)$ is a function of the dispersion of the residual sequence $y(k)$.

In order to determine the weighting factor γ, an interpretation of expression (21) is in order. Note that as $\beta(k)$ becomes small, the pseudo-gain $\alpha(k)$ approaches the conventional Kalman filter gain $K(k)$, and as $\beta(k)$ becomes large, the pseudo-gain approaches unity. Thus for small dispersion of the

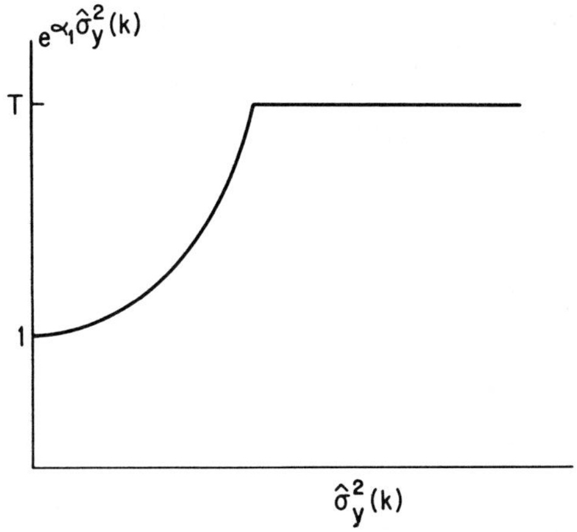

Fig. 3. Influence curve of sample variance for observation residual. (Reproduced from [10] with permission of the ASME.)

residuals, the gain is the optimum Kalman filter gain K(k), while for large dispersion of the residuals no faith is given to the estimation process.

From computer experiments, it was found that a robust weighting function for the propagated error covariance $\overline{P}(k)$ was also required. The modified error covariance was defined as

$$\theta'(k) = \overline{P}(k) + f'[1 - \sigma(k)], \tag{26}$$

where

$$\overline{P}'(k) = \sigma(k)\overline{P}(k). \tag{27}$$

For the scalar case, where $\sigma(k)$ is defined as $e^{-\nu_1 \hat{\sigma}^2(k)}$,

$$\theta'(k) = e^{-\nu_1 \hat{\sigma}_y^2(k)}\,\overline{P}(k) + e^{\alpha_1 \hat{\sigma}_y^2(k)}\,[1 - e^{-\nu_1 \hat{\sigma}_y^2(k)}]. \tag{28}$$

The quantity $e^{\alpha_1 \hat{\sigma}_y^2(k)}$ is limited to some a priori upper bound T. This is illustrated in Fig. 3. Note that for a small dispersion of the residuals [small $\sigma_y(k)$], the modified error covariance $\theta'(k)$ of expression (28) approaches the Kalman propagated error covariance $\overline{P}(k)$. For the other limit, that is, when the dispersion of the residuals is large [large $\sigma_y(k)$], the term $\theta'(k)$ approaches the a priori upper bound T. Note that the estimator gain K(k) is also bounded when the dispersion of the residuals is large or

$$\lim_{\hat{\sigma}_y^2(k) \to \infty} K(k) = \frac{T}{T + R}. \tag{29}$$

The concepts of robust weighting of the Kalman gain and error covariance matrices can be extended to the vector case.

III. ADAPTIVE GAIN MATRIX WEIGHTING

For the case of linear measurements or pseudo-linear measurements, and where the measurements are of all the individual state variables, the ideas presented in expression (21) can be expanded to matrix form to address a vector dynamic case. This matrix formulation is not for the general vector dynamic case, but rather is limited by linear dynamics and complete state vector measurements. This is accomplished with the $\Lambda[K(k)]$ matrix, similar to the $\alpha(k)$ function for the scalar case of (21) [9]. When the measurements are linear, $\Lambda[K(k)]$ will replace the Kalman gain matrix K(k) in the estimation algorithm.

Consider a matrix

$$\Lambda[K(k)] = [I' + EK(k) - F], \tag{30}$$

where I', E, and F are $n \times r$, $n \times n$, and $n \times r$ matrices, respectively. Recall that x(k) and z(k) are n and r vectors, respectively. The individual matrices of (30) can best be explained through example. Consider the linear system of Fig. 4. This system is driven by a deterministic forcing function u(t) and process noise w(t). The output is observed discretely with a sampling sensor system, and these measurements z(k) made by

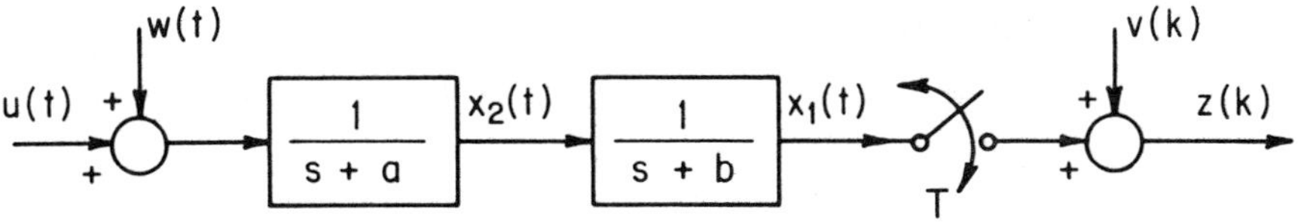

Fig. 4. Linear system example. (Reproduced from [10] with permission of the ASME.)

the sensor are corrupted by a noise sequence v(k). The measurement equation is thus

$$z(k) = x_1(k) + v(k). \tag{31}$$

In matrix form,

$$z(k) = [1 \quad 0]\begin{bmatrix} x_1(k) \\ x_2(k) \end{bmatrix} + v(k). \tag{32}$$

For this example, the I', E, and F matrices are defined as follows:

$$I' = \begin{bmatrix} 1 \\ 0 \end{bmatrix}, \tag{33}$$

$$E = \begin{bmatrix} e^{-\beta(k)} & 0 \\ 0 & 1 \end{bmatrix}, \tag{34}$$

$$F = \begin{bmatrix} e^{-\nu\beta(k)} \\ 0 \end{bmatrix}, \tag{35}$$

where $\beta(k)$ is as defined in relation (25).

IV. ADAPTIVE ERROR COVARIANCE MATRIX WEIGHTING

The equivalent matrix formulation of expression (28) is

$$\theta'(k) = \overline{P}'(k) + F'(I - \Sigma). \tag{36}$$

Again, the individual matrices of (36) can best be explained through application to the linear problem of Fig. 4. The propagated error covariance matrix $\overline{P}(k)$ is partitioned into three separate matrices. The following notation is introduced:

$$\overline{P}(k) = L(k) + D(k) + V(k), \tag{37}$$

where L(k) is a lower triangular matrix with zeros on the diagonal, D(k) is a diagonal matrix, and V(k) is an upper triangular matrix with zeros on the diagonal. The $\overline{P}'(k)$ matrix is defined as

$$\overline{P}'(k) = V(k) + \Sigma D(k) + L(k), \tag{38}$$

where the Σ matrix is a diagonal matrix. The individual elements of the Σ matrix are of the form

$$e^{-\nu_i \hat{\sigma}^2_{y_i}(k)},$$

where $\hat{\sigma}^2_{y_i}(k)$ is a sample variance of the ith observation residual sequence. For the case in which all state variables are measured, all of the diagonal elements of Σ would be of the form

$$e^{-\nu_i \hat{\sigma}^2_{y_i}(k)}.$$

The ith diagonal element Σ_{ii} is

$$e^{-\nu_i \hat{\sigma}^2_{y_i}(k)}.$$

For the problem at hand, where only discrete measurements of the output variable $x_1(k)$ are available,

$$\Sigma = \begin{bmatrix} e^{-\nu_1 \hat{\sigma}^2_{y_1}(k)} & 0 \\ 0 & 1 \end{bmatrix} \tag{39}$$

and

$$F' = \begin{bmatrix} e^{\alpha_1 \hat{\sigma}^2_{y_1}(k)} & 0 \\ 0 & 1 \end{bmatrix}. \tag{40}$$

For a case in which measurements of both state variables were available, the Σ and F' matrices would be, respectively,

$$\Sigma = \begin{bmatrix} e^{-\nu_1 \hat{\sigma}^2_{y_1}(k)} & 0 \\ 0 & e^{-\nu_2 \hat{\sigma}^2_{y_2}(k)} \end{bmatrix}, \tag{41}$$

$$F' = \begin{bmatrix} e^{\alpha_1 \hat{\sigma}^2_{y_1}(k)} & 0 \\ 0 & e^{\alpha_2 \hat{\sigma}^2_{y_2}(k)} \end{bmatrix}. \tag{42}$$

The quantities $e^{\alpha_1 \hat{\sigma}^2_{y_1}(k)}$ and $e^{\alpha_2 \hat{\sigma}^2_{y_2}(k)}$ would be threshold limited in a similar fashion to Fig. 3, with upper bounds of T_1 and T_2, respectively.

Both the adaptive Kalman gain and error covariance weighting procedures were incorporated into the algorithm (4)-(8). It was determined through experimentation that these robust adaptive procedures were most effective if they were activated only after the sample variance of the residual sequence reached a predetermined threshold level. As a rule of thumb, the threshold was taken to be one and one-half times the anticipated value of the standard deviation of the observation noise.

V. ROBUST SMOOTHING

The estimates of the state variables made by the Kalman filter with the modified gain and covariance matrices contain occasional outliers. This is a result of the sampling procedure and the way in which the sample statistics of the residual sequence are utilized to formulate weights for the elements of the gain and covariance matrices. To alleviate the outliers in the state estimates, a robust statistical smoothing procedure was incorporated into the estimation procedure. The robust smoother uses a regression procedure in the following manner. Consider a set λ of m recent estimates of the ith state variable, or

$$\lambda = \Big\{\hat{x}_i(k - m - 1),\ \hat{x}_i(k - m),\ \hat{x}_i(k - m + 1),\ \ldots,\ \hat{x}_i(k)\Big\}. \tag{43}$$

It is desired to find a weighted least squares solution for the straight-line regression fit through the m samples of the estimates of the ith state variable, $\hat{x}_i$, over the discrete temporal interval from k - m - 1 to k. The specific weighted least squares solution for the straight-line regression case (i.e., $\tilde{Y} = \beta_0 + \beta_1\tilde{K} + \tilde{E}$) is given by the formulas

$$\hat{\beta}_1 = \frac{\Sigma_{j=1}^{m}\left[(W_j)(j - \overline{x}')\left(\hat{x}_i(k - m + j) - \overline{Y}'\right)\right]}{\Sigma_{j=1}^{m}\left[(W_j)(j - \overline{x}')^2\right]} \tag{44}$$

and

$$\hat{\beta}_0 = \overline{Y}' - \hat{\beta}_1\overline{x}'. \tag{45}$$

When m, the sample size, is an odd integer, $\overline{x}'$ is defined as

$$\overline{x}' = m - (m + 1)/2 + 1 \tag{46}$$

and $\overline{Y}'$ is defined as

$$\overline{Y}' = \frac{\Sigma_{j=1}^{m}\left[W_j \hat{x}_i(k - m + j)\right]}{\Sigma_{j=1}^{m} W_j} \tag{47}$$

The weighting term W_j that appears in (44) and (47) is called the biweight [19], which is an abbreviation for bisquare weight. Observations (meaning samples of a random variable) are weighted according to the relationship

$$W(e_i) = \begin{cases} \left(1 - e_i^2\right)^2, & |e_i| \leqq 1, \\ 0, & \text{elsewhere,} \end{cases} \tag{48}$$

where

$$e_i = \left(t_i - \hat{t}\right)/cs \tag{49}$$

and t_i is the ith observation, with $\hat{t}$ the estimate of location based on N observations

$$\hat{t} = \frac{\Sigma_{i=1}^{N}[W(e_i)t_i]}{\Sigma_{i=1}^{N} W(e_i)} \tag{50}$$

A robust measure of scale is defined in [20] as

$$s = \frac{\text{(interquartile distance)}}{2(0.6745)}, \tag{51}$$

where the interquartile distance is defined as the third quartile minus the first quartile and thus gives the length of the interval in which the central half of the data falls. For samples that arise from Gaussian distributions, s is an estimate of σ, the standard deviation, and $\hat{t}$ is an estimate of the mean. The value of the constant c is arbitrary.

Discrete values of the weighted least squares solution along the regression line are obtained from the relationship

$$\tilde{Y}(\tilde{k}) = \beta_0 + \beta_1\tilde{k} \quad \text{for} \quad \tilde{k} = 1, 2, \ldots, m, \tag{52}$$

where β_0 and β_1 are defined by (45) and (44). The vertical distances from the regression line of (52) to individual data points at the m discrete times are called residuals and are defined by the relationship

$$r(j) = \hat{x}_i(k - m + j) - \tilde{Y}(j) \quad \text{for} \quad j = 1, 2, \ldots, m. \tag{53}$$

The above expressions of (52) and (53) are used in formulating the robust smoothing procedure illustrated in Fig. 2. This robust procedure is implemented by using the weighted least squares solution of (52) to project m - 1 past values of the estimates (as formulated by the adaptive filter) of the ith state variable up to the present discrete time, t = k. The m - 1 past values of the estimates of the ith state variable,

$$\left\{\hat{x}_i(k - m - 1), \hat{x}_i(k - m), \ldots, \hat{x}_i(k - 1)\right\}, \tag{54}$$

are projected to discrete time t = k and define m values of the random variable $x_j^{\dagger}(k)$, where

$$x_j^{\dagger}(k) = \tilde{Y}(k) + r(j) \quad \text{for} \quad j = 1, 2, \ldots, m. \tag{55}$$

The newly formed random variable $x_j^{\dagger}(k)$ is smoothed by using the relationship

$$\hat{x}_i'(k) = \frac{\Sigma_{j=1}^{m} W_j x_j^{\dagger}(k)}{\Sigma_{j=1}^{m} W_j}, \tag{56}$$

where $\hat{x}_i'(k)$ is the smoothed value of the estimated value of the ith state variable as generated by the modified gain and covariance Kalman filter. A new estimate, at discrete time k + 1, of the ith state variable is generated, $\hat{x}_i(k + 1)$, which is subsequently smoothed by means of the above process; however, the sample space now spans the discrete time interval from time k - m to time k + 1. The sample set λ of (43) is now defined as

$$\lambda = \left\{\hat{x}_i(k - m),\ \hat{x}_i(k - m + 1),\ \ldots,\ \hat{x}_i(k + 1)\right\}. \tag{57}$$

A new weighted least squares solution for the straight-line regression fit through the m samples is found and the process repeats as outlined above.

Note that the solution of the nonlinear relationships of (44), (47), and (56) is obtained from an iterative procedure. Equations (44), (47), and (56) are nonlinear as a result of the bisquare weight function W_j given by the relationship (48).

VI. SIMULATION RESULTS

The system of Fig. 4 (with a = 1 and b = 3) was simulated digitally, and the output $x_1(t)$ was sampled in time and corrupted by discrete sensor noise v(k) with zero mean and a variance of 25. The system was driven only by a deterministic forcing function u(t), which was a pulse with a magnitude of 300 units and a 27-sec duration, as illustrated in Fig. 5. Also shown in Fig. 5 are records of the values of the state variables $x_2(t)$ and $x_1(t)$.

A conventional Kalman filter, without any a priori knowledge of the forcing function u(t) or the time at which the forcing function was applied, was used to process the measurement data z(k). The conventional Kalman filter did not detect the influence of the deterministic forcing function on the state variables, as illustrated in Fig. 6. Since a priori data dictated that there was no process noise, the elements of the Kalman filter gain matrix associated with the observed variables

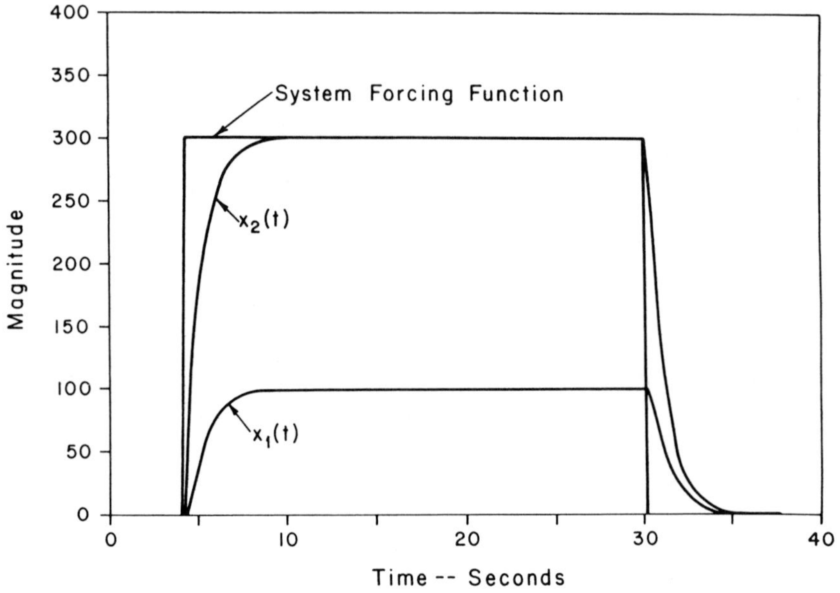

Fig. 5. System forcing function and state variables.

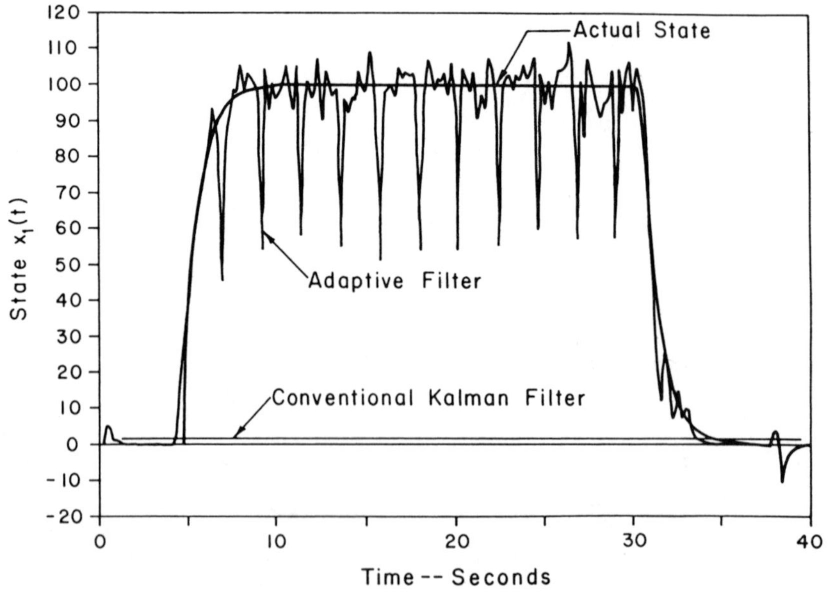

Fig. 6. Estimation of state $x_1(t)$ using an adaptive Kalman filter without robust smoothing compared to the nonadaptive conventional Kalman filter.

approach zero; thus the estimation process has severed itself from the measurement process and ignores new data brought forth by additional measurements.

When the elements of the Kalman filter gain and covariance matrices are weighted by the adaptive procedure outlined above (sample statistics of the residual sequence are used to adapt the respective weights), the filter no longer divorces itself from the measurement process. Additional data brought forth by the measurement process are used to update the estimates of the state variables, as illustrated in Fig. 6. However, since the adaptive procedure uses sample statistics, the estimates contain periodic outliers. The filter will run for a period of time, then monitor the residual sequence to update the adaptive weights. It is this monitoring of the residual sequence to obtain new information that causes the periodic outlier to appear in the estimates. The subsequent processing of the adaptive estimates by a robust smoother reduces the level of mean-square error and the periodic outliers.

The smoothed estimates for $m = 11$ of the measured output state variable $x_1(k)$ are shown in Fig. 7. Figure 8 presents an overlay of the records of Fig. 6 and 7 for 5-sec expanded time interval.

Of prime concern when dealing with stochastic problems in estimation is how well the estimator performs in terms of reducing the level of error in the estimates. The error in the estimate is defined as the difference between the estimate $\hat{x}(k)$ and the actual value $x(k)$. Recall that the estimator of expression (4) minimizes the conditional mean-square error J of

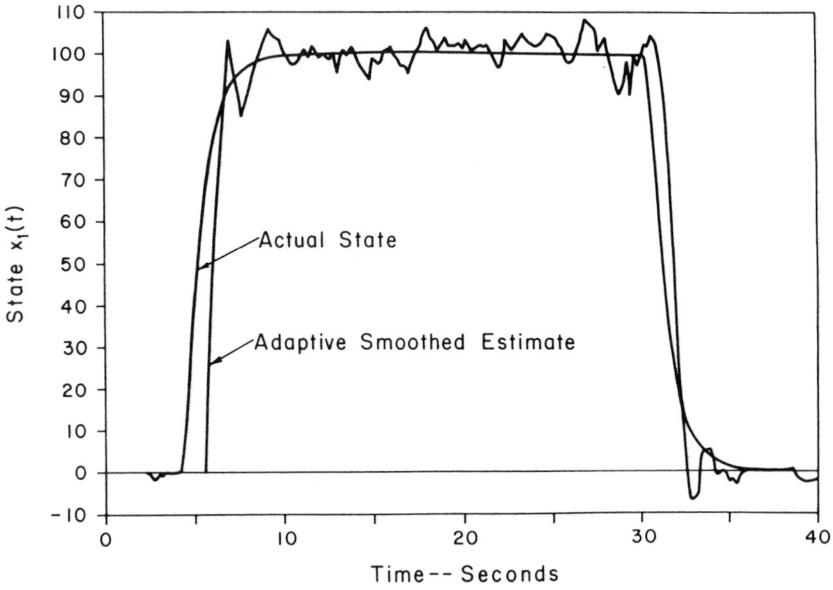

Fig. 7. Estimation of state $x_1(t)$ using the adaptive Kalman filter with robust smoothing.

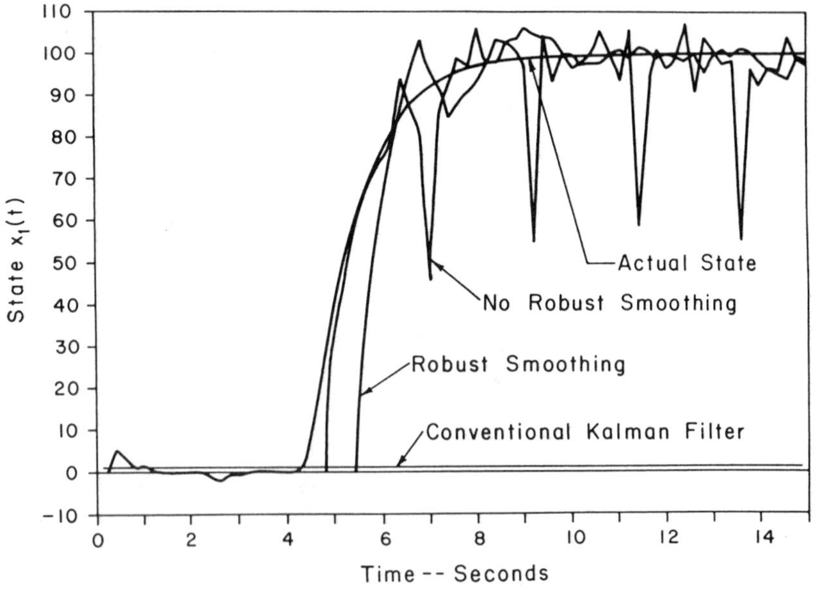

Fig. 8. Comparison of filtering techniques.

expression (3). The performance of the estimator of expression (4) can be evaluated in terms of the covariance propagation as defined in (7).

The performance of an adaptive robust filter, such as the one presented in this section, can be evaluated in a similar manner, that is, through examining the level of mean-square error in the estimates. However, this must be done empirically since a closed-form expression for the covariance propagation has not been developed. A very useful approach is to evaluate the normalized mean-square error noise-to-estimate ratio. This normalized mean-square error ratio, called a "figure of merit," is defined by the expression

$$F_m(k) = 2\,\frac{\hat{\sigma}_z^2(k)/\hat{\sigma}_{\hat{x}}^2(k)}{1 + \hat{\sigma}_z^2(k)/\hat{\sigma}_{\hat{x}}^2(k)}. \tag{58}$$

The quantities $\hat{\sigma}_z^2$ and $\hat{\sigma}_{\hat{x}}^2$ in the above expression are moving-window recursive sample variances as defined by expression (B.11) of Appendix B. Examination of expression (58) shows that when the two variances are equal, the figure of merit $F_m(k)$ is unity. In this situation the estimator is not improving the estimates over the sensor measurements. However, when $\hat{\sigma}_z^2(k) \gg \hat{\sigma}_{\hat{x}}^2(k)$, the figure of merit approaches two, indicating that the estimator has reduced the errors contributed by the measurement process to zero. The other extreme is the case of a poor or unacceptable estimator, when $\hat{\sigma}_{\hat{x}}^2(k) \gg \hat{\sigma}_z^2(k)$, in which case the figure of merit $F_m(k)$ approaches zero.

The figure of merit, as defined in expression (57), is very useful from an empirical standpoint for evaluating adaptive estimator performance. Figure 9 presents figure-of-merit data for the estimates of $x_1(k)$ made by the adaptive estimator with

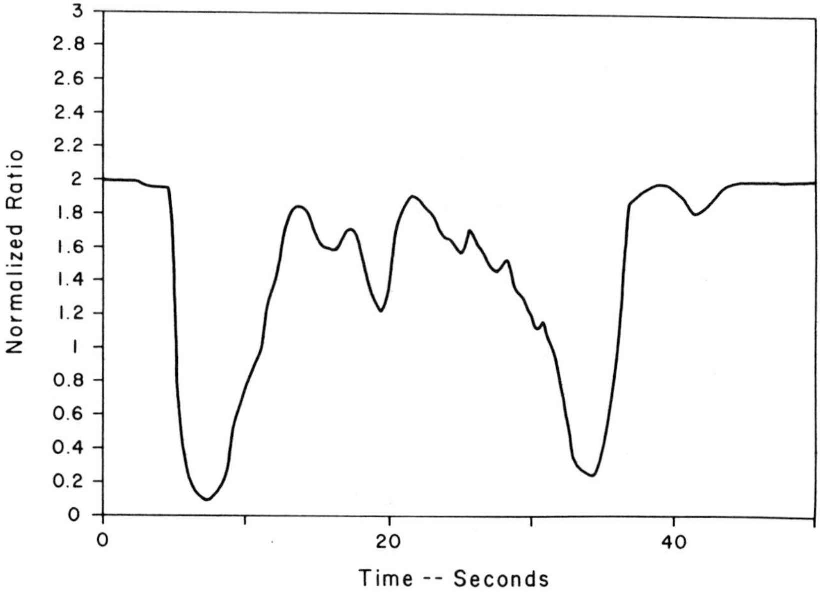

Fig. 9. Time variation of normalized mean-square error ratio.

robust smoothing. Through examination of Fig. 9 it is noted that during the transition periods of the system forcing function the estimator is performing in the "poor" category. However, during the time period when the system forcing function is at its maximum (throughout the period of the pulse), the estimator is performing respectably at reducing sensor measurement errors below the sensor noise level. Similarly, during the periods when the system is not being forced, the data of Fig. 9 show very good estimator performance, with the figure of merit approaching two.

VII. CONCLUSIONS

State variable estimation with incomplete system information (noise statistics, forcing functions, and system dynamics) is not a problem to which there are clearcut solutions. The article addresses the cases of missing information about the system deterministic forcing functions and noise statistics. The concepts presented relative to this particular problem address the limited class of linear system dynamics with associated linear measurements. Nonlinear system dynamics with associated linear or nonlinear measurements, however, are not precluded.

Estimates of the state variable using the adaptive process for the system during the periods when the system is not being forced are relatively close to those of the conventional Kalman filter for congruent periods, but there is some increase in mean-square error because the adaptive estimator is no longer optimal. During periods when the system is being forced a vast improvement, as compared with those estimates of the conventional Kalman filter, is realized with the adaptive gain, covariance weight, and associated robust smoothing procedure. The estimates derived with the adaptive procedure during the periods of system forcing do, however, contain a considerable level of mean-square error. This seems to be a prevailing shortfall of adaptive estimation procedures. The trade-off is knowledge of the deterministic forcing functions versus high mean-square estimate error in the absence of that knowledge.

APPENDIX A: ROBUST ESTIMATORS OF STATISTICS

Myers and Tapley [7] proposed a method of estimating an unknown forcing constant $u(k) = u$ and any bias r in the measurement noise. These estimates are

$$\hat{u}(k) = \frac{1}{N_s} \sum_{j=k-N_s+1}^{k} f(j) \tag{A.1}$$

and

$$\hat{r}(k) = \frac{1}{N_z} \sum_{j=k-N_z+1}^{k} y(j), \tag{A.2}$$

where

$$f(k) = \hat{x}(k) - A\hat{x}(k - 1) \tag{A.3}$$

and

$$y(k) = z(k) - A\hat{x}(k - 1) - \hat{u}(k - 1). \tag{A.4}$$

The covariance estimators were given to be

$$\hat{Q} = \frac{1}{N_s - 1} \sum_{j=k-N_s+1}^{k} (f(j) - \hat{u}(k))(f(j) - \hat{u}(k))^T - \frac{1}{N_s} \sum_{j=k-N_s+1}^{k} (A\hat{P}(j - 1)A^T - \hat{P}(j)) \tag{A.5}$$

and

$$\hat{R} = \frac{1}{N_z - 1} \sum_{j=k-N_z+1}^{k} (y(j) - \hat{r}(k))(y(j) - \hat{r}(k))^T - \frac{1}{N_z} \sum_{j=k-N_z+1}^{k} H(A\hat{P}(j - 1)A^T + \hat{Q}(j - 1))H^T. \tag{A.6}$$

In the following it is stated that an equally valid approach gives a different form of $\hat{Q}$. It also will be shown that $\hat{u}$ and $\hat{r}$ cannot be estimated simultaneously. Lastly are proposed estimators of u, r, Q, and R that are robust to contaminated Gaussian noise distributions and other heavy-tailed densities.

Groutage [9] rederived the Q covariance estimator (A.5) but used $f'(k) = \hat{x}(k) - A\hat{x}(k - 1) - \hat{u}(k - 1) = \hat{x}(k) - \overline{x}(k)$ to do so.

The analysis in [10] shows that the correct $\hat{Q}$ is

$$Q = \frac{1}{N_s - 1} \sum_{j=k-N_s+1}^{k} (f'(j) - \hat{u}'(k))(f'(j) - \hat{u}'(k))^T$$

$$- \frac{1}{N_s} \sum_{j=k-N_s+1}^{k} (A\hat{P}(j - 1)A^T - \hat{P}(j)), \qquad (A.7)$$

where $\hat{u}'(k)$ is the average of the $f'(j)$. Averaging large numbers of simulations has shown that this estimator leads to accurate results, $\hat{Q}$ going negative only rarely. (Myers and Tapley used $|\hat{Q}|$ to overcome negative estimates. This study chooses instead to use the most recent positive estimate.) Note that (A.1) must still be used to estimate u.

Next the instability of the variance estimators demonstrated in [7] and remarked on in other literature is considered here. A flowgraph showing system states, the Kalman filter, and estimators for r and u is given in Fig. 10. It is clear that the

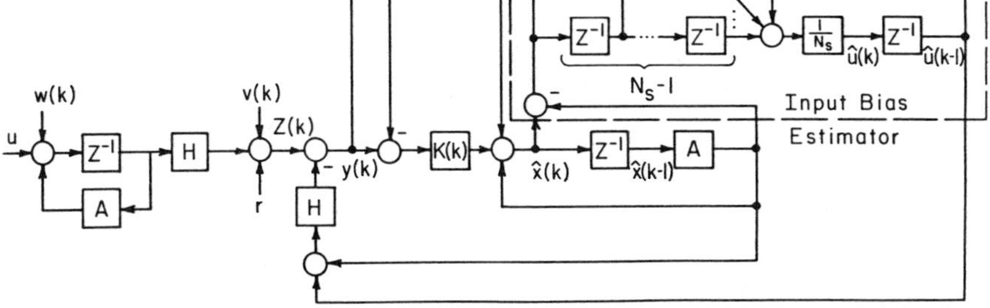

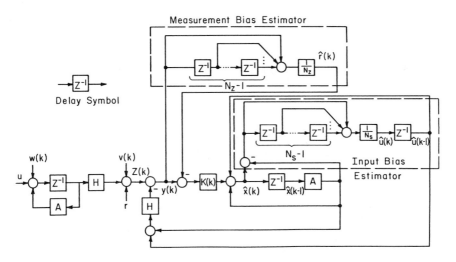

Fig. 10. Myers and Tapley algorithm for estimation of input bias u and measurement bias r.

constants appearing as inputs at the summing junction whose output is z(k) cannot be distinguished after their addition. Analysis of the iterations* in the scalar case verifies the proportioning of any one of the biases into the two estimators. This erroneous division of the biases leads to erroneous $\hat{Q}$ and $\hat{R}$, which in turn contaminate the Kalman gain and $\hat{P}$ matrices. A solution to this difficulty has not yet been proposed other than to know a priori one of the biases, particularly r, since we are interested in unknown step inputs.

The last item of concern in this appendix is the exposition of robust variance and covariance estimators for the elements of $\hat{Q}$ and $\hat{R}$. Expressions of the type (A.1) for sample mean and (A.7) for sample covariance may be made robust with the use of medians and bisquare weights as follows, as in [19].

Let the robust estimator of bias of a stationary independent, identically distributed sequence s_i be the median over N samples,

$$m_s = \text{med}\{s_i\}_N, \tag{A.8}$$

and let the variance of the sequence be estimated by

$$\hat{\sigma}_s^2 = \frac{N\,\Sigma_i'(s_i - m_s)^2\left(1 - \mu_i^2\right)^4}{\left[\Sigma_i'\left(1 - \mu_i^2\right)\left(1 - 5\mu_i^2\right)\right]\left[-1 + \Sigma_i\left(1 - \mu_i^2\right)\left(1 - 5\mu_i^2\right)\right]}, \tag{A.9}$$

where

$$\mu_i = \frac{s_i - m_s}{\eta\ \text{med}\{|s_i - m_s|\}_N}. \tag{A.10}$$

The authors acknowledge Alireza Moghaddamjoo for this work and also for the verification of the formula (A.7). Details of these analyses are to be given in his Ph.D. thesis, Electrical Engineering Department, University of Wyoming, and also in a technical report for 1984 to the Office of Naval Research, Contract No. N0014-82-K-0048.

The weights $\left(1 - \mu_i^2\right)^2$ are called bisquare weights or biweights, Σ' indicates summation over all N elements in the data window except those for which $\mu_i^2 > 1$, and η is a number such as six or nine. Using the relationship $\sigma_{12}^2 = \sigma_1\sigma_2\rho_{12}$, the covariance matrix P for the $n \times 1$ random variable X can be written as

$$P = \begin{bmatrix} \sigma_i^2 & \rho_{12}\sigma_1\sigma_2 & \cdots & \rho_{1n}\sigma_1\sigma_n \\ \rho_{12}\sigma_1\sigma_2 & \sigma_2^2 & \cdots & \rho_{2n}\sigma_2\sigma_n \\ \vdots & \vdots & & \\ \rho_{1n}\sigma_1\sigma_n & \rho_{2n}\sigma_2\sigma_n & \cdots & \sigma_n^2 \end{bmatrix}. \tag{A.11}$$

The covariance matrix as defined by expression (A.11) can be expressed in terms of the standard deviation matrix D and the correlation matrix R, where

$$D \triangleq \begin{bmatrix} \sigma_1 & 0 & \cdots & 0 \\ 0 & \sigma_2 & & \\ \vdots & & \ddots & \vdots \\ 0 & & & \sigma_n \end{bmatrix} \tag{A.12}$$

and

$$R \triangleq \begin{bmatrix} 1 & \rho_{12} & \cdots & \rho_{1n} \\ \rho_{21} & 1 & & \\ \vdots & & \ddots & \\ \rho_{n1} & & & 1 \end{bmatrix}. \tag{A.13}$$

The covariance matrix is then defined as

$$P \triangleq DRD. \tag{A.14}$$

A robust estimate of P can be formulated based on robust estimates of D and R. Expression (A.9) can be used to establish a robust estimate of D. There now only remains the task of defining a robust estimate of R to complete the development of a robust covariance estimator. This task can be approached from a variety of viewpoints. Spearman's rank correlation coefficient, as presented in [21], is defined as

$$\hat{\rho}_{ij} = 1 - \frac{6 \ \Sigma_{k=1}^{N} \ d_i(k)}{N(N^2 - 1)}, \tag{A.15}$$

where

$$d_i(k) = X_i(k) - X_j(k).$$

Similarly, estimators of ρ_{ij} can be formulated using an order statistic approach, as presented in [22]. If the pairs $X_i(k)$ and $X_j(k)$ are ordered by their $X_i(k)$ variates, where the rth-ordered $X_i(k)$ is denoted $X_{i(r)}$, then the $X_j(k)$ variate associated with the rth-ordered $X_i(k)$ is denoted $X_{j[r]}$ and termed the concomitant of the rth-order statistic. Using these definitions the following two estimators for the ρ_{ij} correlation coefficients are presented:

$$\hat{\rho}_{ij} = \frac{\overline{X}'_{j[K]} - \overline{X}_{j[K]}}{\overline{X}'_{j(K)} - \overline{X}_{j(K)}} \tag{A.16}$$

and

$$\hat{\rho}'_{ij} = \frac{1}{2} \frac{\overline{X}'_{j[K]} - \overline{X}_{j[K]}}{\overline{X}'_{j(K)} - \overline{X}_{j(K)}} + \frac{\overline{X}'_{i[K]} - \overline{X}_{i[K]}}{\overline{X}'_{j(K)} - \overline{X}_{i(K)}}, \tag{A.17}$$

where

$$X'_{j(K)} = \frac{1}{K} \sum_{\ell=1}^{K} X_{j(N+1-\ell)}, \tag{A.18}$$

$$\overline{X}_{j(K)} = \frac{1}{K} \sum_{\ell=1}^{K} X_{j(\ell)}, \tag{A.19}$$

$$\overline{X}'_{j[K]} = \frac{1}{K} \sum_{\ell=1}^{K} X_{j[N+1-\ell]}, \tag{A.20}$$

and

$$\bar{X}_{j[K]} = \frac{1}{K} \sum_{\ell=1}^{K} X_{j[\ell]}. \tag{A.21}$$

Robust estimators for R are obtained by replacing the ρ_{ij} terms of expression (A.13) with either of the estimators of (A.15), (A.16), or (A.17). Note that the covariance estimators of expressions (A.5) and (A.6) for Q and R, respectively, contain two sample estimators of statistics. The first term in either (A.5) or (A.6) is an estimator of covariance, whereas the second term for either expression is an estimator of location. The sample estimate of covariance can be estimated using the robust estimator of (A.14), whereas the sample estimate of location can be obtained using a robust estimate of location such as the median, that is, the estimator or expression (A.8).

Extensive simulation has been conducted to evaluate sample statistics of variates s and t by robust methods described above, where s and t are sequences of random events. The covariance of sequences s and t may be found using the relationship $\sigma_{st}^2 = \sigma_s \sigma_t \rho_{st}$, where the correlation coefficient ρ_{st} is estimated by any of the expressions (A.15), (A.16), or (A.17). The simulations were conducted on Gaussian and contaminated Gaussian sequences, for which the probability density of s is defined by

$$p(s) = (1 - \epsilon)N(0, 1) + \epsilon N(0, \sigma_2), \tag{A.22}$$

and t, the covariate, is defined by the relationship

$$t = 0.57s + w, \tag{A.23}$$

where

$$p(w) = N(0, \sigma_w). \tag{A.24}$$

In the simulations σ_2 and σ_w were varied with such that always $\sigma_t^2 = 7.0$ and $\sigma_{st} = 2.0$. Typical results with sample size N = 15 showed that the normalized mean-square errors of robust estimators gave improvements by factors of 10 to 100 over conventional estimators of both variances and covariances.

It is clear that in heavy-tailed noise densities these robust estimators for $\hat{u}$, $\hat{r}$, $\hat{Q}$, and $\hat{R}$ will improve the conventional adaptive estimators of Myers and Tapley, with sample medians replacing sample means and biweighted squares and robust covariances replacing the linear sample covariance elements.

The findings of this appendix should yield considerable improvements in practical adaptive Kalman filtering problems.

APPENDIX B: DERIVATION OF RECURSIVE ESTIMATORS FOR SAMPLE STATISTICS

Presented in this appendix are the derivations of the recursive estimators for the sample mean and sample variance based on N_ℓ observations.

RECURSIVE SAMPLE MEAN

The expression for the recursive estimator for the sample mean at time t_{k-1} is

$$\hat{n}(k-1) = \frac{1}{N_\ell} \sum_{j=k-N_\ell}^{k-1} n(j). \tag{B.1}$$

The expression for the recursive estimator for the sample mean at time t_k is

$$\hat{n}(k) = \frac{1}{N_\ell} \sum_{j=k-N_\ell+1}^{k} n(j). \tag{B.2}$$

Equation (B.1) is subtracted from (B.2) to give

$$\hat{n}(k) - \hat{n}(k-1) = \frac{1}{N_\ell}\left[\sum_{j=k-N_\ell+1}^{k} n(j) - \sum_{j=k-N_\ell}^{k-1} n(j)\right]. \qquad (B.3)$$

When the terms under the summation of (B.3) are expanded out and appropriate cancellations take place, the recursive estimator for the sample mean at time t_k is given as

$$\hat{n}(k) = \hat{n}(k-1) + \frac{1}{N_\ell}[n(k) - n(k-N_\ell)]. \qquad (B.4)$$

RECURSIVE SAMPLE VARIANCE

The expression for the recursive estimator for the sample variance at time t_k is

$$\hat{\sigma}_n^2(k) = \frac{1}{N_\ell - 1}\sum_{j=k-N_\ell+1}^{k}[n(j) - \hat{n}(k)]^2. \qquad (B.5)$$

Equation (B.5) is rewritten as

$$\hat{\sigma}_n^2(k) = \frac{1}{N_\ell - 1}\left\{[n(k) - \hat{n}(k)]^2 - \left[\hat{n}(k-N_\ell) - \hat{n}(k)\right]^2 + \sum_{j=k-N_\ell}^{k-1}[n(j) - \hat{n}(k)]^2\right\}. \qquad (B.6)$$

When the expression for the sample mean (B.4) is substituted into the summation term of (B.6), it can be rewritten as

$$\hat{\sigma}_n^2(k) = \frac{1}{N_\ell - 1}\left\{[n(k) - \hat{n}(k)]^2 - \left[n(k-N_\ell) - \hat{n}(k)\right]^2 + \sum_{j=k-N_\ell}^{k-1}\left[(n(j) - \hat{n}(k-1)) - \frac{1}{N_\ell}(n(k) - n(k-N_\ell))\right]^2\right\}. \qquad (B.7)$$

Expanding the terms under the summation gives

$$\hat{\sigma}_n^2(k) = \frac{1}{N_\ell - 1}\Bigg\{[n(k) - \hat{n}(k)]^2 - \Big[n(k - N_\ell) - \hat{n}(k)\Big]^2$$

$$+ \sum_{j=k-N_\ell}^{k-1} \Bigg[(n(j) - \hat{n}(k - 1))^2 - \frac{2}{N_\ell}(n(j) - \hat{n}(k - 1))$$

$$\times \Big(n(k) - \hat{n}(k - N_\ell)\Big)$$

$$+ \frac{1}{N_\ell^2}\Big(n(k) - \hat{n}(k - N_\ell)\Big)^2\Bigg]\Bigg\}. \qquad (B.8)$$

The cross terms under the summation of the expression (B.8) can be shown to be zero; thus,

$$\hat{\sigma}_n^2(k) = \frac{1}{N_\ell - 1}\Bigg\{[n(k) - \hat{n}(k)]^2 - \Big[n(k - N_\ell) - \hat{n}(k)\Big]^2$$

$$+ \sum_{j=k-N_\ell}^{k-1} \Bigg[(n(j) - \hat{n}(k - 1))^2$$

$$+ \frac{1}{N_\ell^2}\Big(n(k) - \hat{n}(k - N_\ell)\Big)^2\Bigg]\Bigg\}. \qquad (B.9)$$

Note that by definition

$$\hat{\sigma}_n^2(k - 1) = \sum_{j=k-N_\ell}^{k-1} [n(j) - \hat{n}(k - 1)]^2; \qquad (B.10)$$

thus, the recursive estimator for the sample variance is

$$\hat{\sigma}_n^2(k) = \hat{\sigma}_n^2(k - 1) + \frac{1}{N_\ell - 1}\Bigg\{[n(k) - \hat{n}(k)]^2$$

$$- [n(k - N_\ell) - \hat{n}(k)]^2$$

$$+ \frac{1}{N_\ell^2}\Big[n(k) - \hat{n}(k - N_\ell)\Big]^2\Bigg\}. \qquad (B.11)$$

REFERENCES

1. K. F. GAUSS, "Theory of Motion of the Heavenly Bodies," Dover, New York, 1963.

2. N. WIENER, "The Extrapolation, Interpolation, and Smoothing of Stationary Time Series with Engineering Applications," MIT Press, Cambridge, Massachusetts, 1949.

3. R. E. KALMAN, "A New Approach to Linear Filtering and Prediction Theory," *ASME J. Basic Eng. Ser. D 82*, 35-45 (1960).

4. R. E. KALMAN and R. S. BUCY, "New Results in Linear Filtering and Prediction Theory," *ASME J. Basic Eng. Ser. D 83*, 95-108 (1961).

5. Y. C. HO and R. C. K. LEE, "A Bayesian Approach to Problems in Stochastic Estimation and Control," *IEEE Trans. Autom. Control AC-9*, No. 5, 333-339 (1969).

6. R. G. JACQUOT, "Modern Digital Control Systems," Dekker, New York, 1981.

7. K. A. MYERS and B. D. TAPLEY, "Adaptive Sequential Estimation with Unknown Noise Statistics," *IEEE Trans. Autom. Control AC-21*, No. 4, 520-523 (1976).

8. R. J. FITZGERALD, "Divergence of the Kalman Filter," *IEEE Trans. Autom. Control AC-16*, No. 6, 736-747, December 1971.

9. F. D. GROUTAGE, "Adaptive Robust Sequential Estimation with Application to Tracking a Maneuvering Target," Ph.D. Dissertation, Department of Electrical Engineering, University of Wyoming, May, 1982.

10. F. D. GROUTAGE, R. G. JACQUOT, and D. E. SMITH, "Adaptive State Estimation Using Robust Smoothing," *Trans. ASME J. Dyn. Syst. Meas. Control 106*, No. 4, 335-341, December 1984.

11. R. A. SINGER, "Estimating Optimal Tracking Filter Performance for Manned Maneuvering Target, " *IEEE Trans. Aerosp. Electron. Syst. AES-6*, No. 4, 473-483, July, 1970.

12. R. F. BERG, "Estimation and Prediction for Maneuvering Target Trajectories," *IEEE Trans. Autom. Control AC-28*, No. 3, 294-303, March, 1983.

13. R. K. MEHRA, "On the Identification of Variances and Adaptive Kalman Filtering," *IEEE Trans. Autom. Control AC-15*, No. 2, 175-184, April, 1970.

14. T. KAILATH, "An Innovations Approach to Least Squares Estimation, Part I: Linear Filtering in Additive White Noise," *IEE Trans. Autom. Control AC-13*, No. 6, 646-654, December 1968.

15. A. H. JAZWINSKI, "Adaptive Filtering," *Automatica 5*, No. 4, 475-485, June, 1969.

16. J. R. OLMSTEAD and L. C. GOHEEN, "A Multiunit Relative Localization and Tracking Algorithm for OTH Targeting," Technical Report NWRC-TR-22, Naval Warfare Research Center.

17. P. J. BICKEL, "One-Step Huber Estimates in the Linear Model," *J. Am. Stat. Assoc. 70*, No. 350, 428-433 (1975).

18. P. J. HUBER, "Robust Estimation of a Location Parameter," *Ann. Math. Stat. 35*, No. 1, 73-101 (1964).

19. F. MOSTELLER and J. W. TUKEY, "Data Analysis and Regression," Addison-Wesley, Reading, Massachusetts, 1977.

20. R. L. LAUNER and G. N. WILKINSON, "Robustness in Statistics," Academic Press, New York, 1979.

21. W. MENDENHALL, "Introduction to Probability and Statistics," Wadsworth, Berlmont, California, 1971.

22. H. A. DAVID, "Order Statistics," Wiley, New York, 1981.

Coordinate Selection Issues in the Order Reduction of Linear Systems*

ANDREI L. DORAN

The Aeorspace Corporation
El Segundo, California

I. INTRODUCTION

In many engineering investigations it is desirable or necessary to reduce the complexity of the equations used to describe a physical system or component. Simplification can be achieved, for example, by reducing the number of equations or by transforming the set of equations into a form containing fewer parameters. Both aspects will be addressed here.

The reduction in the number of equations is usually called order reduction, or model reduction, and forms the main focus of the present work. The reduction in the number of parameters is usually referred to as parameter simplification. Parameter simplification will also be addressed in this article, but only to the extent that it helps the order reduction investigation.

The practical problems motivating model reduction are varied [1]. One such problem of considerable current interest is the control of large space structures [2-6]. Tight pointing and attitude requirements, combined with limited-size on-board

This work was supported by the Aerospace Sponsored Research Program.

processors, necessitate high-quality reduced-order models. Beyond any particular application, the model reduction field has grown into a well-established body of research. Being able to intelligently discard some of the system equations has several obvious benefits. Repetitive simulations become easier and cheaper to perform. Salient features of the system, previously hidden in a mass of detail, may be revealed. And, controllers may be designed for the reduced model, since most currently available control design methods only work on small-dimension systems. The robustness of such controllers is another important issue.

A brief survey of the model reduction literature is provided in the next section, backed by a more thorough review in [1]. A common way to accomplish model reduction is by coordinate truncation, that is, by discarding a subset of the original coordinates. Skelton's method of cost decomposition provides a framework for determining which coordinates to retain and which to discard. A feature of this approach is that the properties of the reduced model may be strongly influenced by the choice of coordinates used to describe the original model. This article explores some of the relationships between the original coordinate selection and properties of the reduced model obtained by cost decomposition.

The class of "cost uncoupled" coordinates is the main focus of the article. Conditions are obtained under which stability of the reduced model can be assured given that the original model is stable. The quality of reduced models obtained from several different but equivalent descriptions of the large model are compared, and a class of coordinate transformations that do not affect the quality of the reduced model is identified.

Parameter simplification is the second aspect pertaining to the reduction of complexity. This article explores parameter simplifications that can be effected on linear systems in preparation for model reduction by cost decomposition. It is of interest to evaluate the suitability of different coordinate selections for use in the full-order system prior to model reduction. This evaluation cannot be performed on a general, full-parameter original system, even if it is of small dimension. Parameter-simplifying transformations are identified that do not affect the model reduction quality. That is, the model reduction quality based on the parameter-simplified system is the same as that based on the full-parameter system. A numerical example shows how systematic use of parameter simplification enables a complete evaluation of coordinate selections for order reduction.

II. LITERATURE SURVEY

Model reduction research is limited to linear systems because only for such systems is order a measure of complexity [7]. But many practical examples, including large space structures, lead to linear models. The literature pertaining to order reduction of linear systems is vast, as can be seen from the lengthy reference list of Genesio and Milanese's much quoted article [8]. Other surveys can be found in [1,9-13]. Shorter reviews can be found in the introductions of [14,15]. In the review of the large space structure control problem [4], model reduction is also treated briefly. Between all the above surveys one can obtain an appreciation of the field. However, most of these reviews are not recent. This is why some important new contributions, like Skelton's ideas on cost decomposition [2,16,17] or Moore's "balanced coordinates" [18,191 are only mentioned in [1]. Another interesting and quite recent development is Hyland's optimal projection approach to model and controller reduction [20-22].

To discuss the large body of model reduction literature, it is convenient to classify it into several types and then focus on some representative articles. Chen [14] has one such classification, and Bosley and Lees [9] have another. Skelton [16] suggests three categories of model reduction; polynomial approximations, component truncations, and parameter optimization methods, but does not survey any articles. These three categories will be used here because they seem to make a more fundamental distinction than whether the system is originally expressed in the time or frequency domain, in state space, or in transfer matrix form. This choice is not really important, since some papers have elements of two or all three categories and they would fit more neatly in another classification.

The polynomial approximation methods generally use low-order transfer functions whose coefficients are sought such that the output matches the output of the large-order original system as closely as possible. One well-known example in this class is the determination of simple models from experimental frequency responses using Bode plots, by asymptotes and break points. Another example of a simple polynomial approximation technique is based on discarding the less important time constants from the full-order transfer function to obtain a reduced-order transfer function. This method is related to the retention of dominant modes to be mentioned among component truncations.

Most polynomial approximation methods are more sophisticated variations of the above. One popular method is based on matching moments of the reduced-order transfer function to those of the original transfer function [23-26]. A variation of this approach matches "cumulants" [25]. The moment matching method leads to quick convergence of the steady-state response. Convergence of the transient response is improved by also matching the first few Markov parameters [10,27].

Related to moment matching is the fundamental polynomial approximation method using Padé approximants. This method investigates basic convergence issues pertinent to all polynomial approximations and has received a good deal of attention [27-31]. The method can be extended directly to multi-input multi-output (MIMO) systems [27]. One problem with all polynomial approximations, shared by the Padé approximants, is that preservation of stability (or instability) of the original model is not guaranteed in the reduced model [27,28,30]. To alleviate this situation, the reduced-system poles can be determined from a Routh table formed from the original system [28,30].

Another popular subclass of polynomial approximations is the method of continued fractions [32-36]. In a comparison of several methods [13], continued fractions gave good results. This may be explained by [34], where it is shown that the continued-fraction method gives a "best" approximation in some sense. The method can be extended to MIMO cases [14]. It can be shown that the continued-fraction reduced-order model is a Padé approximant [26,27]. The relationship between continued fractions and Padé approximants is explored in more detail in

[37]. In [33] an improvement is given to avoid obtaining unstable (stable) reduced models from stable (unstable) original models.

Other polynomial approximation methods select the coefficients of the reduced-order model that minimize an error functional between the outputs of the full and reduced systems [38-40]. These papers will be mentioned among the parameter optimization category. A more extended discussion of polynomial approximation techniques, as well as of the other approaches covered in this survey, can be found in [1].

The second class of methods considered here are the component truncations. They usually apply to models expressed in state-space form and obtain the reduced-order model by retaining a subset of the original system. Therefore the reduced-model coefficients are more constrained than in the parameter optimization case discussed later, where they can be freely synthesized. As such, the reduced models obtained from component truncations may only give "suboptimal" results, but many of these approaches are simple in concept and can be applied to very large-order systems. In addition to discarding rows and columns of the system matrices, various criteria are used to guide these component truncations, and many methods use coordinate transformations to perform the truncation in "more favorable" spaces.

The simplest version among component truncations diagonalizes the state matrix (assuming a full set of eigenvectors) and retains the cominant modes. The methods mentioned below are variations on this theme. One nice feature of these approaches as compared to the polynomial approximations is that the reduced models produced here always preserve the stability of the full-order models, since their eigenvalues are similar. On the other hand, the steady-state response of the reduced model is usually different from that of the full-order model, especially in the simpler approaches.

Davison [41] retains not only the dominant eigenvalues but also the eigenvectors associated with them. In order words, the eigenvectors of the r-order reduced model have elements in the same ratios as the first r elements of the original eigenvectors. Later [42], a diagonal transformation on the reduced state was proposed to ensure correct steady-state response. A different improvement, also ensuring correct reduced-model steady-state response, appeared in [43]. An interesting method for systems without numerator dynamics (zeros) is based on successively eliminating the largest eigenvalues [44]. For better matching of all phases of the response (initial, intermediate, and steady state), a combination of three reduced models is proposed in [45].

Mitra [46,47] explicitly recognized model truncation as a projection, thus anticipating Hyland [20-22]. Mitra also suggested retaining the components with the highest entry in a diagnonalized controllability matrix, thus preceding Moore's "balanced" approach [18,19]. Another simple but effective method [48] discards the components with high eigenvalues, but only after setting them to their constant steady-state values (which are reached fast due to their small time constants).

Yet another popular approach is based on singular perturbation methods [49-52], wherein the original system is divided into slow and fast modes. A "small" parameter is identified, and the system is rearranged so that the small parameter multiplies the derivatives of the fast modes. The fast modes are discarded, but after being set to their quickly reached steady-state values (very similar to [48]).

Skelton's cost decomposition method [2,16,17] and Moore's balanced coordinates approach [18,19] will be described in more detail in the following sections. Other well-known component truncation methods include Kalman's minimum realizations [53-55] and Aoki's "aggregation" [56].

The third category of model reduction techniques, called parameter optimization here, usually employs a numerical iteration scheme for synthesizing the elements of the reduced model that minimizes an error function of the difference between full- and reduced-order outputs. Sometimes the conditions to be satisfied by an optimum reduced model can be nicely written as concise matrix equations [13,57,58], but these are nonlinear and the solutions have to be numerical.

Anderson's optimal projection using a discretized version of the continuous system [59] is described in detail in [1]. This method tries to match the trajectories of the full- and reduced-order models, as in the moment matching technique of polynomial approximations. Similar optimal projection can be applied to models in transfer function form instead of state-space form [38-40].

An important group of papers is based on explicit optimality conditions. A single-input single-output (SISO) system in transfer function form is considered in [7], but most ensuing papers deal with MIMO systems in state-space form. A detailed review of Wilson's much quoted work [13,57,58] is given in [1]. Nice optimality conditions are derived for the reduced model in state-space form, and a numerical procedure for synthesizing the coefficients is given. It is also shown that the optimum reduced-model steady-state error is orthogonal to the reduced output [1,58]. The method is extended from systems driven by impulses or white noise to systems with general polynomial inputs [57]. A number of good papers [15,60,61] are close to Wilson's. Reference [15] is able to express the optimality conditions as scalar equations by focusing on SISO systems. A simplification is obtained in [60] by treating the inputs one at a time, and in [61] by arbitrarily fixing the coefficients in either the input or output matrices. Finally, in [40], the usual squared error function is replaced with the maximum difference between the full- and reduced-order outputs.

This survey has not been all encompassing, and more detail can be found in [1]. If the field of linear system reduction seems like a potpourri of ad hoc techniques, this is because it is less than 25 years old. If there is one unifying characteristic, it is that practically all methods are input dependent. There is no universally best model reduction scheme [7,18]. The work to be presented next deals with rational ways of reducing state-space representations and belongs in the component truncation category described above.

III. MODEL REDUCTION BY COST DECOMPOSITION

The cost decomposition, or component cost analysis (CCA), approach to model reduction was introduced by Skelton and is described in [1,16,17]. A brief summary follows.

Consider the system

$$\begin{aligned} \dot{\underline{x}} &= A\underline{x} + D\underline{w}, \quad E[\underline{w}(t)] = 0, \\ \underline{y} &= C\underline{x}, \quad E\left[\underline{w}(t)\underline{w}^T(\tau)\right] = W\delta(t - \tau), \end{aligned} \tag{1}$$

where $\underline{x} \in R^n$, $\underline{w} \in R^k$, $\underline{y} \in R^p$, and the stationary convariance matrix W is positive definite. The matrices A, D, and C are constant, and the eigenvalues of A have negative real parts. The pairs (A, D) and (A, C) are disturbable and observable [62], respectively. A scalar measure of the system response (also referred to as the total response function) may be defined as

$$V \triangleq \lim_{t\to\infty} E\|\underline{y}\|^2 = \lim_{t\to\infty} E\left[\underline{w}^T D^T K D \underline{w}\right] = \mathrm{tr}[KG], \tag{2}$$

where K is the symmetric, positive-definite, unique solution of the Lyapunov equation

$$A^T K + KA + C^T C = 0 \tag{3}$$

and G is the "input matrix"

$$G \triangleq DWD^T. \tag{4}$$

The n-coordinate state vector $\underline{x}$ can be partitioned into N components:

$$\begin{aligned} \underline{x}^T &= \Big\{x_1 x_2, \ldots, x_{n_1} \mid x_{n_1+1}, \ldots, x_{n_1+n_2} \mid \cdots \\ &\qquad \mid x_r, \ldots, x_{r+n_i} \mid \ldots, x_n\Big\} \\ &\triangleq \Big\{\underline{X}_1^T \quad \underline{X}_2^T \quad \cdots \quad \underline{X}_i^T \quad \cdots \quad \underline{X}_N^T\Big\}, \qquad r = 1 + \sum_{j=1}^{i-1} n_j. \end{aligned} \tag{5}$$

Written in partitioned form using these components, system (1) becomes

$$\begin{aligned} \dot{\underline{X}}_i &= \sum_{j=1}^{N} A_{ij}\underline{X}_j + D_i\underline{w}, \qquad \underline{X}_i \in R^{n_i}, \\ \underline{y} &= \sum_{j=1}^{N} C_j\underline{X}_j, \qquad \sum_{i=1}^{N} n_i = n. \end{aligned} \tag{6}$$

Define the input on component i as

$$\hat{\underline{w}}_1 \triangleq D_i\underline{w}, \qquad \hat{\underline{w}} = D\underline{w}. \tag{7}$$

Then the cost function defined in (2) becomes

$$V = \lim_{t\to\infty} E\left[\sum_{i=1}^{N}\sum_{j=1}^{N} \hat{\underline{w}}_i^T K_{ij} \hat{\underline{w}}_j\right], \tag{8}$$

where K_{ij} is the ij subpartition of matrix K from (3). The "value" or "cost" of a component is defined as

$$V_i \triangleq \lim_{t\to\infty} E\left[\frac{1}{2}\,\frac{\partial\|\underline{y}\|^2}{\partial\hat{\underline{w}}_i}\,\hat{\underline{w}}_i\right] = E\sum_{j=1}^{N} \hat{\underline{w}}_j^T K_{ij} \hat{\underline{w}}_i, \tag{9}$$

where $\partial V/\partial\hat{\underline{w}}_i$ is the row vector of the partials of V with respect to the elements of $\hat{\underline{w}}_i$. Thus the ith component cost depends on both the sensitivity of the total cost to the ith input itself. Substituting (8) and (9) gives, after a little algebra,

$$V_i = \mathrm{tr}[KG]_{ij} = \sum_{j=1}^{N} \mathrm{tr}\, K_{ij} G_{ij}^T = \sum_{j=1}^{N} V_{ij}. \tag{10}$$

V_i may be regarded as a measure of the degree to which $\underline{x}_i$ contributes to the total system response. A basic assumption of the method of cost decomposition is that if the output of the reduced model is to closely match the output of the large model, components $\underline{x}_i$ that make a large contribution to V should be retained.

Assuming that the components of the state vector are renumbered such that they appear in order of decreasing cost,

$$V_1 \geq V_2 \geq \cdots V_R > V_{R+1} \geq \cdots \geq V_N, \tag{11}$$

the first R components are retained in the reduced model. Partitioning the system of (1) into "retained" and "truncated" parts, one obtains

$$\begin{Bmatrix} \dot{\underline{x}}_R \\ \dot{\underline{x}}_T \end{Bmatrix} = \begin{bmatrix} A_R & A_{RT} \\ A_{TR} & A_T \end{bmatrix} \begin{Bmatrix} \underline{x}_R \\ \underline{x}_T \end{Bmatrix} + \begin{bmatrix} D_R \\ D_T \end{bmatrix} \underline{w}, \qquad \underline{y} = [C_R \quad C_T] \begin{Bmatrix} \underline{x}_R \\ \underline{x}_T \end{Bmatrix}. \tag{12}$$

The reduced model is simply

$$\dot{\hat{\underline{x}}}_R = A_R\hat{\underline{x}}_R + D_R\underline{w}, \qquad \hat{\underline{y}} = C_R\hat{\underline{x}}_R. \tag{13}$$

The purpose of CCA model reduction is to keep $\underline{y}$ "close" to $\hat{\underline{y}}$. The performance measure, or the model error index (MEI), used is

$$\text{MEI} = \frac{\Delta V}{V} \triangleq \frac{\lim_{t\to\infty} E\|\underline{y} - \hat{\underline{y}}\|^2}{\lim_{t\to\infty} E\|\underline{y}\|^2}. \tag{14}$$

δV can be evaluated conveniently from an aggregate system containing both the original and reduced models with an output of $\underline{y} - \hat{\underline{y}}$:

$$\dot{\underline{x}}' = \begin{Bmatrix} \dot{\underline{x}}_R \\ \dot{\underline{x}}_T \\ \dot{\tilde{\underline{x}}}_R \end{Bmatrix} = \begin{bmatrix} A_R & A_{RT} & 0 \\ A_{TR} & A_T & 0 \\ 0 & A_{RT} & A_R \end{bmatrix} \begin{Bmatrix} \underline{x}_R \\ \underline{x}_T \\ \tilde{\underline{x}}_R \end{Bmatrix} + \begin{bmatrix} D_R \\ D_T \\ 0 \end{bmatrix} \underline{w} \triangleq A'\underline{x}' + D'\underline{w},$$

$$\underline{y}' = \underline{y} - \hat{\underline{y}} = [0 \quad C_T \quad C_R]\underline{x}' \triangleq C'\underline{x}', \qquad \tilde{\underline{x}}_R = \underline{x}_R - \hat{\underline{x}}_R. \tag{15}$$

δV is the response function of the primed system in the same way that V of (2) was the response function of the system in (1). It then follows that

$$\delta V = \text{tr } K'G', \qquad A'^T K' + K'A' + C'^T C' = 0,$$
$$G' = D'WD'^T. \tag{16}$$

Some features of the method just described are worth noting.

(a) The procedure applies with little or no modification to alternative models, such as when the input vector $\underline{w}$ is deterministic instead of stochastic, when the output of interest is only over a finite time span, or even when the system matrices A, D, C, W are time varying [1,17]. If the input is a set of random initial conditions, instead of the white noise, then the order reduction problem is entirely equivalent [13,16].

(b) The approach provides a way of deciding which coordinates to keep and which to discard. But this choice, as well as the truncation quality as reflected in the MEI, is influenced by the selection of the original coordinates in $\underline{x}$ of (1) in which the truncation takes place. This is an important part of the work to be described shortly.

(c) In general, there is no guarantee that the reduced model is stable when the original model is stable. There are some coordinate choices wherein stability of the reduced model can be assured. One such choice is when the dynamic matrix is diagonal (or "modal" [1,16]). Another choice is when the cost is "uncoupled" in a special way, and this will be elaborated on in a later section of this article.

(d) The component costs V_i of (10) can be negative. When this happens, the magnitude of V_i should be used in (11) as a measure of the component's value. In the "cost uncoupled" coordinates to be described, all component costs are positive.

(e) The cost, or value, of a component V_i reflects the three classical criteria for the importance of a component: the time constant, the input coupling, and the output coupling [1,16]. This is seen implicitly in (10), and in the special case of coordinates that are both dynamically uncoupled and cost uncoupled the component cost simply becomes the product of the three classical criteria mentioned above: $V_i = \tau_i \|\underline{c}_i\|^2 \|\underline{d}_i\|_W^2$, where $\underline{c}_i$, $\underline{d}_i$ are the ith column and row of C, D.

IV. COORDINATE SELECTION

Consider a nonsingular coordinate transformation

$$\underline{x} = T\underline{\eta}. \tag{17}$$

In the new coordinates, $\underline{\eta}$, system (1) becomes

$$\begin{aligned} \dot{\underline{\eta}} &= T^{-1}AT\underline{\eta} + T^{-1}D\underline{w} \triangleq \bar{A}\underline{\eta} + \bar{D}\underline{w}, \\ \underline{y} &= CT\underline{\eta} \triangleq \bar{C}\underline{\eta}. \end{aligned} \tag{18}$$

The total response function is the same as before (however, the component costs will, in general, be different):

$$V = \lim_{t\to\infty} E\|y\|^2 = \operatorname{tr} \bar{K}\bar{G} = \sum_{i=1}^{N} V_i, \tag{19}$$

where

$$\begin{aligned} &\bar{A}^T\bar{K} + \bar{K}\bar{A} + \bar{C}^T\bar{C} = 0, \quad \bar{K} = T^TKT, \\ &G = \bar{D}W\bar{D}^T, \quad \bar{G} = T^{-1}GT^{-T}, \end{aligned} \tag{20}$$

$$\bar{V}_i = \sum_{j=1}^{N} \operatorname{tr} K_{ij}G_{ij}^T \triangleq \sum_{j=1}^{N} \bar{V}_{ij}. \tag{21}$$

In the development to follow, the phrase "coordinate selection" will refer to the selection of T in (17). Some choices of T proposed in [16] are

(i) select T such that $\bar{A}$ is block diagonal (dynamic uncoupling);

(ii) select T such that $\bar{C}_T = 0$ (output uncoupling);

(iii) select T such that $\bar{G}$ is diagonal (disturbance uncoupling);

(iv) select T such that $\bar{V}_{ij} = 0$, $\forall i \neq j$ (cost uncoupling).

The first choice above is well known and is the basis of many component truncation schemes based on frequency. Some elaboration on the other three choices is useful.

The output uncoupling transformation

$$\overline{C} = CT = \left[\overline{C}_R \quad \overline{C}_T\right] = \left[\overline{C}_R \quad 0\right] \tag{22}$$

can be implemented only if ℓ, the number of columns in C_T, obeys the relationship [1]

$$\ell \leq n - \text{rank}(C). \tag{23}$$

Thus, output uncoupling cannot be applied when fewer states than independent outputs are retained.

Another property of output uncoupling is that components corresponding to zero columns in C are not necessarily the components of lowest cost (though they happen to be the low-cost components for the examples considered in [16]). In cases where the cost of components with nonzero columns in C is lower than components with zero columns in C, then truncating the latter components is incompatible with the philosophy of the cost decomposition approach, which is based on low-cost component truncation. Because of these properties and also because of the generally unimpressive performance of this approach as reported in [16], output uncoupling is not considered further in this article. Recent results reported in [63], however, indicate the output uncoupling in the context or model reduction by aggregation has useful properties for the design of closed-loop controllers for structural systems.

Both of the transformations required by choices (iii) and (iv) uncouple the cost. In fact, choice (iii) is a sufficient condition for (iv). The necessary condition for cost uncoupling is obtained directly from (21):

$$\text{tr } K_{ij}G_{IJ}^T = 0, \quad \forall i \neq j. \tag{24}$$

However, the sufficient condition $K_{ij}G_{ij}^T = 0$ is more amenable to implementation by coordinate selection. There are just four ways in which $K_{ij}G_{ij}^T = 0$. Either of the matrices in the product can be zero, or neither, or both. If $G_{ij} = 0$, such as in (iii), or if $K_{ij} = 0$ it will be said that the cost is "singly" uncoupled, or just "uncoupled." If both $K_{ij} = 0$ and $G_{ij} = 0$, the cost will be called "doubly" uncoupled.

Another set of cost-uncoupled coordinates are the "balanced coordinates" of B. C. Moore [18]. To obtain balanced coordinates, the matrix T of (17) is chosen so as to diagonalize K in the congruence transformation of (20) and also to diagonalize the state covariance matrix X. In terms of the original coordinates, the state covariance matrix X is defined as the expected value of $\underline{x}\underline{x}^T$. X may be found as the unique, positive-semidefinite solution of the Lyapunov equation

$$AX + XA^T + G = 0$$

(see [62]). If the pair (A, D) is disturbable, then X is positive definite. For the new coordinates, the state covariance matrix X satisfies a similar equation:

$$\bar{A}\bar{X} + \bar{X}\bar{A}^T + \bar{G} = 0. \tag{26}$$

$\bar{X}$ and X are related by the equation

$$\bar{X} = T^{-1}XT^{-T}. \tag{27}$$

In [18] and [19] the transformation T not only diagonalizes K and X, it also makes the diagonalized versions of these matrices equal. From the point of view of model reduction, this last requirement will be shown to be unimportant.

Cost-uncoupled coordinates have several appealing features. For example, $\bar{V}_{ij} = 0$ for $i \neq j$, so that $\bar{V}_i = \bar{V}_{ii}$ [see (21)]. This means that each component cost is a function only of the ith state or component and contains no cross coupling. The doubly uncoupled coordinates seem especially attractive in this respect. Another nice property of cost-uncoupled coordinates is that the component costs are always positive. This can be seen from (8), which can be expanded with (10), and clearly $V_{ii} \geq 0$

$$V = \sum_{i=1}^{N} V_i = \sum_{i=1}^{N} \sum_{i=1}^{N} V_{ij}, \qquad V_{ii} = \lim_{t\to\infty} E\left(\underline{\hat{w}}_i^T K \underline{\hat{w}}_i\right). \tag{28}$$

The balanced coordinates have their own intuitive appeal for use in model reduction. To see this, note that the observability and disturbability matrices W_o and W_c defined as

$$W_o \triangleq \int_0^{\infty} e^{A^T t} C^T C e^{At}\, dt,$$
$$W_c \triangleq \int_0^{\infty} e^{At} DWD^T e^{A^T t}\, dt \tag{29}$$

(see [62]) satisfy the Lyapunov equations (3) and (25), respectively. Therefore

$$W_o = K, \quad W_c = X. \tag{30}$$

In the transformed coordinates (denoted by overbars), the matrices are diagonal:

$$\overline{W}_o = \overline{W}_c = \Sigma. \tag{31}$$

Thus the balanced coordinates are coordinates that diagonalize the observability and disturbability matrices. In [18] and [19], the reduced model is obtained by truncating the components with the smallest diagonal entries in Σ. This is reasonable since a small entry in Σ implies that a component is not very observable or disturbable. However, this is not the procedure used here. The intent of the present article is to evaluate various coordinate selections for truncation by component cost analysis. In the numerical example shown in the next section, the truncation (even in balanced coordinates) is made by component cost and not by examining the sizes of the elements in Σ. (It was found, however, that components with high cost also had large entries in Σ.)

The appeal of the above properties is largely intuitive. Another property, however, concerns an important practical consideration: stability. When the coordinates describing the larger order model are not cost uncoupled, it is possible to obtain a reduced model that is unstable even when the full model is asymptotically stable. When certain cost-uncoupled coordinates are employed, this undesirable result will not occur. This can be formalized in the following theorem.

Theorem 1

If the original system of (1) is asymptotically stable and the coordinate transformation of (17) is chosen such that $\overline{K}$ in (20) has the form

$$\overline{K} = \begin{bmatrix} \overline{K}_R & 0 \\ 0 & \overline{K}_T \end{bmatrix},$$

then the reduced model is stable.

Proof. Rewriting (2) in partitioned form yields

$$\begin{bmatrix} \overline{A}_R^T & \overline{A}_{TR}^T \\ \overline{A}_{RT}^T & \overline{A}_T^T \end{bmatrix}\begin{bmatrix} \overline{K}_R & 0 \\ 0 & \overline{K}_T \end{bmatrix} + \begin{bmatrix} \overline{K}_R & 0 \\ 0 & \overline{K}_T \end{bmatrix}\begin{bmatrix} \overline{A}_R & A_{RT} \\ \overline{A}_{TR} & \overline{A}_T \end{bmatrix}$$

$$= -\begin{bmatrix} \overline{C}_R^T\overline{C}_R & \overline{C}_R^T\overline{C}_T \\ \overline{C}_T^T\overline{C}_R & \overline{C}_T^T\overline{C}_T \end{bmatrix}. \tag{32}$$

The reduced-model matrices are A_R, C_R, and D_R, and for the cost of the reduced model the Lyapunov equation is

$$\overline{A}_R^T\overline{K}_R + \overline{K}_R\overline{A}_R + \overline{C}_R^T\overline{C}_R = 0. \tag{33}$$

Since $\overline{A}$ is asymptotically stable, and (A, C) form an observable pair, $\overline{K}$ is positive definite. Since $\overline{K}$ is uncoupled, each block $\overline{K}_R$ and $\overline{K}_T$ must be positive definite. Now the pair $\left(\overline{A}_R, \overline{C}_R\right)$ either is observable, or not. Both these cases are covered by [19, Lemma 3.1]:

(a) If $\left(\overline{A}_R, \overline{C}_R\right)$ is an observable pair, then by applying [19, Lemma 3.1] to (27), the positive definiteness of K_R implies that A_R is asymptotically stable.

(b) If $\left(\overline{A}_R, \overline{C}_R\right)$ is not an observable pair, then, by [19, Lemma 3.1], $\overline{A}_R$ is only "stable." The eigenvalues of $\overline{A}_R$ that are not in the open left-half plane are on the imaginary axis and correspond to unobservable modes.

Since in no case can eigenvalues of $\overline{A}_R$ be in the open right-half plane, the theorem is proven. Note that if desired, the unobservable part of the reduced system can be truncated with no penalty in MEI. If this is done, the final reduced model is asymptotically stable.

V. NUMERICAL EXAMPLE

To examine the performance of the various coordinate selections, a simple numerical example was considered. In this example, a two-coordinate model was reduced to one coordinate. The initial system was taken from [16]:

$$\dot{\underline{x}} = \begin{bmatrix} -1 & 0 \\ 0 & -10 \end{bmatrix} \underline{x} + \begin{bmatrix} 1 & 1 \\ 70 & 1 \end{bmatrix} \underline{w}, \quad E[\underline{w}(t)] = \underline{0},$$

$$y = (1 \; - 0.2)\underline{x}, \quad E\left[\underline{w}(t)\underline{w}^T(\tau)\right] = \delta(t - \tau). \tag{34}$$

The component costs were computed for each coordinate selection, and then the MEI was calculated when (a) the low-cost coordinate was truncated and (b) the high-cost coordinate was truncated. (Truncating the high-cost coordinate is counter to the philosophy of component cost analysis. It was included here to examine the difference in MEI for high-cost and low-cost truncation.) In addition to the results based on truncation, a reduced model obtained by parameter optimization was considered. A general first-order model was assumed, and the parameter values for that model were obtained by numerical search with the objective of minimizing the MEI. The results are tabulated in Table 1.

Table 1. Comparison of Truncation Schemes

Coordinate selection	MEI for low-cost component truncation	MEI for high-cost component truncation
Original	0.1217	1.1924
Cost uncoupled		
Disturbance uncoupled $K_{12} \neq 0$ $G_{12} = 0$	0.1053	0.9409
K uncoupled $K_{12} = 0$ $G_{12} \neq 0$	0.1043	0.7959
Doubly uncoupled $K_{12} = 0$ $G_{12} = 0$	0.0770[a]	1.1118
Balanced coordinates $K_{12} = 0$ $G_{12} \neq 0$ $X_{12} = 0$	0.2567	0.3394
Dynamically uncoupled	0.1217	1.1924
Parameter optimal (Not a truncation)	0.0727	

[a]Lowest MEI.

In this example the MEI for low-cost component truncation is lower than the MEI for high-cost component truncation in all coordinate choices. Among the various coordinate selections, the doubly cost uncoupled fares best, with MEI not far from the parameter optimal result.

After obtaining the results displayed in Table 1, a large number of runs were made in which the various truncation schemes were used on many different models. The results sometimes differed from those in Table 1. In particular, it was found that

(1) sometimes truncation of the low-cost component leads to a higher MEI than truncation of the high-cost component;

(2) no single truncation scheme listed in Table 1 (excluding parameter optimization) yielded the lowest MEI in all cases;

(3) each of the truncation schemes that uncouple the cost yielded the lowest MEI in some cases.

It would be impractical to present all of the data generated. However, the appendix contains examples illustrating some of the points made above. The first example is one in which truncating the high-cost component is best. The second example is one in which disturbance uncoupling is best, and the third example is one in which K uncoupling is best. The fourth example is one in which truncation in balanced coordinates is best.

Although a number of exceptions were encountered, truncation of the low-cost components was found to yield the lowest MEI in the overwhelming majority of cases. Of the various cost-uncoupling truncation schemes, both double-uncoupled coordinates and balanced coordinates were found to produce the lowest MEI much more often then either G uncoupling or K uncoupling. Double uncoupling was best somewhat more often than balanced coordinates, but no clear preference was established between these two schemes. These results stimulated the interest in undertaking a more comprehensive evaluation to determine in which regions of the parameter space do the various coordinate selections perform best. The parameter simplification described in Section VIII permitted the performance of just such an evaluation in Section IX.

The results just described raise the question of when can one be assured that retention of the high-cost components will lead to a lower MEI than retention of the low-cost components. One result along these lines appears in [16], where it is shown that if the coordinates are both dynamically and cost uncoupled, then truncating the low-cost components is sure to yield the lowest MEI. The usefulness of this result is limited, however, because it is not, in general, possible to uncouple both the dynamics and the cost. Furthermore, even if it were possible to achieve both dynamic and cost uncoupling for a particular model, there is no guarantee that a different selection of coordinates might not yield an even lower MEI.

A second result on this question may be found in [17]. This result is based on the use of a different truncation quality index:

$$J \triangleq \frac{\lim_{t\to\infty} E\|\underline{y}\|^2 - \lim_{t\to\infty} E\|\underline{y}\|^2}{\lim_{t\to\infty} E\|\underline{y}\|^2} = \frac{V - V_R}{V}. \tag{35}$$

For open-loop systems such as those under consideration here, the index J of (35) is less meaningful than the MEI of (14) because J = 0 does not imply that $|\underline{y} - \hat{\underline{y}}|$ is small. The use of J instead of the MEI, however, does make it much easier to study the effects of truncation analytically. For example, it is shown in [17] that truncating the low-cost components in doubly cost-uncoupled coordinates always leads to the smallest value of J. A modified form of J may be more meaningful than the MEI for systems with feedback, where J may represent a scalar measure of performance and control effort.

VI. EFFECTS OF SKEWING ON THE MEI

In the previous two sections, transformation matrices T that diagonalize A or diagonalize K or G or both are considered. Once the objective of the transformation is chosen, the matrix T is determined uniquely except for the normalization of its columns. The results in this section will indicate that the value of the MEI is not affected by this normalization.

Consider the coordinate transformation of (17) partitioned by columns

$$T = [\underline{t}_1 \mid \underline{t}_2 \mid \cdots \mid \underline{t}_n]. \tag{36}$$

In the discussion to follow, the term "skewing T" will mean multiplying the columns of T by scalars:

$$\text{skewed } T \triangleq \tilde{T} \triangleq [s_1\underline{t}_1 \mid s_2\underline{t}_2 \mid \cdots \mid s_n\underline{t}_n] = TS, \tag{37}$$

where

$$S = \begin{bmatrix} s_1 & & \cdots & 0 \\ 0 & s_2 & \cdots & 0 \\ \vdots & \vdots & \ddots & \vdots \\ 0 & 0 & \cdots & s_n \end{bmatrix}. \tag{38}$$

If T has any uncoupling properties, so does $\tilde{T}$. For example, assume A of (18) is diagonal. Then, using $\tilde{T}$ gives

$$\tilde{A} = \tilde{T}^{-1}A\tilde{T} = S^{-1}T^{-1}ATS = S^{-1}\bar{A}S,$$

$$\begin{bmatrix} 1/s_1 & 0 & \cdots & 0 \\ 0 & 1/s_2 & & 0 \\ \vdots & \vdots & \cdots & \vdots \\ 0 & 0 & \cdots & 1/s_n \end{bmatrix} \begin{bmatrix} \bar{a}_{11} & 0 & \cdots & 0 \\ 0 & \bar{a}_{22} & & 0 \\ \vdots & \vdots & \cdots & \vdots \\ 0 & 0 & \cdots & \bar{a}_{nn} \end{bmatrix} \begin{bmatrix} s_1 & 0 & \cdots & 0 \\ 0 & s_2 & & 0 \\ \vdots & \vdots & \cdots & \vdots \\ 0 & 0 & \cdots & s_n \end{bmatrix} = \bar{A}. \tag{39}$$

Similarly, if K or G of (20) is diagonal,

$$\tilde{K} = \tilde{T}^T K \tilde{T} = S^T T^T KTS = S\bar{K}S = \bar{K}S^2 = \text{diagonal}, \tag{40}$$

$$\tilde{G} = \tilde{T}^{-1} G T^{-T} = S^{-1} T^{-1} G T^{-T} S^{-T} = S^{-1} \bar{G} S^{-T} = \bar{G} S^{-2} = \text{diagonal}. \tag{41}$$

It is to be expected that the uncoupling properties of T would be preserved in $\tilde{T}$. When T diagonalizes a matrix by a similarity or a congruence transformation, it is usually an eigenvector matrix, and eigenvectors are only determined within a scalar factor. However, it is also true that the transformations

$$\underline{x} = T\underline{\eta} \quad \text{and} \quad \underline{x} = \tilde{T}\underline{\xi} \tag{42}$$

lead to different vectors $\underline{\eta}$ and $\underline{\xi}$. Thus, it is surprising that truncating in either $\underline{\eta}$ or $\underline{\xi}$ leads to a reduced model with the same MEI! Before stating this result in the form of a theorem, it is useful to note from (37) and (42) that $\underline{\eta}$ and $\underline{\xi}$ satisfy the equation

$$\underline{\eta} = S\underline{\xi}, \tag{43}$$

where S is a diagonal matrix. It is this relationship between $\underline{\eta}$ and $\underline{\xi}$ that forms the basis for this theorem.

Theorem 2

Suppose two sets of coordinates $\underline{\eta}$ and $\underline{\xi}$ satisfy (43) for a diagonal matrix S. If cost decomposition techniques are used to obtain a reduced model, then the MEI for the reduced model will be the same regardless of whether $\underline{\eta}$ or $\underline{\xi}$ is used to describe the full-order model.

Proof. Partitioned in terms of retained and truncated coordinates, (43) can be written as

$$\begin{Bmatrix} \underline{\eta}_R \\ \underline{\eta}_T \end{Bmatrix} = \begin{bmatrix} S_R & 0 \\ 0 & S_T \end{bmatrix} \begin{Bmatrix} \underline{\zeta}_R \\ \underline{\zeta}_T \end{Bmatrix}. \tag{44}$$

Write (18) using $\underline{\eta}$ and $\underline{\xi}$ partitioned as in (44):

$$\begin{aligned} \begin{Bmatrix} \dot{\underline{\eta}}_R \\ \dot{\underline{\eta}}_T \end{Bmatrix} &= \begin{bmatrix} \bar{A}_R & \bar{A}_{RT} \\ \bar{A}_{TR} & \bar{A}_T \end{bmatrix} \begin{Bmatrix} \underline{\eta}_R \\ \underline{\eta}_T \end{Bmatrix} + \begin{bmatrix} \bar{D}_R \\ \bar{D}_T \end{bmatrix} \underline{w}, \\ \underline{y} &= [\bar{C}_R \quad \bar{C}_T] \begin{Bmatrix} \underline{\eta}_R \\ \underline{\eta}_T \end{Bmatrix}, \end{aligned} \tag{45}$$

$$\begin{aligned} \begin{Bmatrix} \dot{\underline{\xi}}_R \\ \dot{\underline{\xi}}_T \end{Bmatrix} &= \begin{bmatrix} \tilde{A}_R & \tilde{A}_{RT} \\ \tilde{A}_{TR} & \tilde{A}_T \end{bmatrix} \begin{Bmatrix} \underline{\xi}_R \\ \underline{\xi}_T \end{Bmatrix} + \begin{bmatrix} \tilde{D}_R \\ \tilde{D}_T \end{bmatrix} \underline{w}, \\ \underline{y} &= [\tilde{C}_R \quad \tilde{C}_T] \begin{Bmatrix} \underline{\xi}_R \\ \underline{\xi}_T \end{Bmatrix}. \end{aligned} \tag{46}$$

The cost of the system is the same whether written in terms of $\underline{\eta}$ using (45) or $\underline{\xi}$ using (46). From (20) and (43) one obtains

$$\tilde{V} = \text{tr } \tilde{K}\tilde{G} = \text{tr } S^T\bar{K}SS^{-1}\bar{G}S^{-T} = \text{tr } \overline{KGS^{-T}S^T} = \text{tr } \overline{KG} = \bar{V} = V. \tag{47}$$

More importantly, the component cost hierarchy is also preserved. Letting $F = KG$, and using $S^T = S$, one has

$$\begin{aligned} V_i &= \text{tr}[SFS^{-1}]_{ii} = \text{tr}\left\{[s_i][\delta_{ij}][s_j^{-1}]\right\}_{ii} = \text{tr}\left\{[s_i\delta_{ij}s_j^{-1}]\right\}_{ii} \\ &= \sum_{i=r}^{r+n_i} s_i\delta_{ii}s_i^{-1} = \sum_{i=r}^{r+n_i} \delta_{ii} = \text{tr}[F]_{ii} = \bar{V}_i \end{aligned}$$

[see (5) for r].

The matrices of (44-46) are related by

$$\begin{aligned} \tilde{A}_R &= S_R^{-1}\bar{A}_R S_R, & \tilde{A}_{RT} &= S_R^{-1}\bar{A}_{RT} S_T, \\ \tilde{A}_{TR} &= S_T^{-1}\bar{A}_{TR} S_R, & \tilde{A}_T &= S_T^{-1}\bar{A}_T S_T, \\ \tilde{D}_R &= S_R^{-1}\bar{D}_R, & \tilde{D}_T &= S_T^{-1}\bar{D}_T, \\ \tilde{C}_R &= \bar{C}_R S_R, & \tilde{C}_T &= \bar{C}_T S_T. \end{aligned} \tag{48}$$

Writing out the expanded system for $\underline{\xi}'$ from which the index $\delta\tilde{V}$ is calculated,

$$\tilde{A}' = \begin{bmatrix} \tilde{A}_R & \tilde{A}_{RT} & 0 \\ \tilde{A}_{TR} & \tilde{A}_T & 0 \\ 0 & \tilde{A}_{RT} & \tilde{A}_R \end{bmatrix} = \begin{bmatrix} S_R^{-1}\bar{A}_R S_R & S_R^{-1}\bar{A}_{RT} S_T & 0 \\ S_T^{-1}\bar{A}_{TR} S_R & S_T^{-1}\bar{A}_T S_T & 0 \\ 0 & S_R^{-1}\bar{A}_{RT} S_T & S_R^{-1}\bar{A}_R S_R \end{bmatrix}$$

$$= \begin{bmatrix} S_R^{-1} & 0 & 0 \\ 0 & S_T^{-1} & 0 \\ 0 & 0 & S_R^{-1} \end{bmatrix} \begin{bmatrix} \bar{A} & \bar{A}_{RT} & 0 \\ \bar{A}_{TR} & \bar{A}_T & 0 \\ 0 & \bar{A}_{RT} & \bar{A}_R \end{bmatrix} \begin{bmatrix} S_R & & 0 \\ 0 & S_T & 0 \\ 0 & 0 & S_R \end{bmatrix} \triangleq S'^{-1}\bar{A}'S'. \tag{49}$$

Similarly,

$$\tilde{D}' = S'^{-1}D', \qquad \tilde{C}' = \bar{C}'S'. \tag{50}$$

Therefore

$$\underline{\eta}' = S'\underline{\xi}' \tag{51}$$

and

$$\delta\tilde{V} = \text{tr } \tilde{K}'\tilde{G}' = \text{tr } S'^T\bar{K}'S'S'^{-1}\bar{G}S'^{-T} = \text{tr } \bar{K}'\bar{G}' = \delta\bar{V}. \tag{52}$$

This completes the proof. The crux of the proof lies in the fact that $\underline{\eta}$ and $\underline{\xi}$ are related by a transformation simple enough (diagonal) in (43), so that the expanded vectors $\underline{\eta}'$ and $\underline{\xi}'$ are also related by a diagonal coordinate transformation. A simple extension follows directly from the fact that skewing T is equivalent to having $\underline{\eta}$ and $\underline{\xi}$ satisfy (43).

Theorem 3

For two sets of coordinates $\underline{\eta}$ and $\underline{\xi}$ satisfying (43), the truncation index of (35) yields the same value. (The proof is similar to the one described above and will not be given here.)

One interesting application of the above results concerns the balanced coordinates introduced previously. In [8], the use of balanced coordinates is shown to reduce ill-conditioning with respect to inversion of the $\overline{W}_o$ and $\overline{W}_c$ matrices. The properties of these coordinates for model reduction are also examined. They are found to be intuitively appealing and in a numerical example they lead to a truncation yielding a low MEI. In view of Theorem 1 it is clear that any lowering of the MEI for reduced models obtained with these coordinates must be due to simultaneously diagonalizing $\overline{W}_o$ and $\overline{W}_c$ and not to setting them equal becuase truncations of identical quality would be obtained by skewing, where

$$\tilde{W}_o = S^2\Sigma, \quad \tilde{W}_c = S^{-2}\Sigma. \tag{53}$$

VII. PERMISSIBLE PARAMETER-SIMPLIFYING TRANSFORMATIONS

When a set of equations is transformed to one containing fewer parameters, some properties of the original system may be lost. This section explores transformations that do not change the MEI upon model reduction by cost decomposition.

Suppose a transformation is applied to the system in (1) to produce a new system:

$$\begin{aligned} \dot{\underline{z}} &= F\underline{z} + G\underline{q}, \quad E[\underline{q}(t)] = 0, \\ \underline{y} &= H\underline{z}, \quad E\left[\underline{q}(t)\underline{q}^T(\tau)\right] = Q\delta(t - \tau). \end{aligned} \tag{54}$$

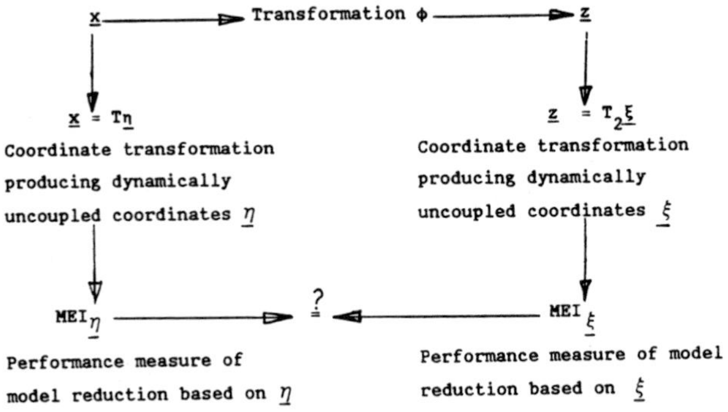

Fig. 1. Definition of permissible transformation ϕ.

The question is whether the performance measure of model reduction in any coordinate selection based on $\underline{x}$ of (1), MEI_x, is the same as the performance measure of model reduction in the similar coordinate selection based on $\underline{z}$, MEI_z.

A graphical illustration of the question may help visualize the issue (Fig. 1).

If MEI_η is equal to MEI_ξ, for all coordinate selections of interest, then one can perform a comparative evaluation of model reduction in the coordinate selections based on $\underline{z}$ instead of on their equivalents based on $\underline{x}$. It may be very advantageous to do this evaluation based on $\underline{z}$, if ϕ is such that the system in (25) contains fewer parameters than system (1).

Several transformations were found to be in the permissible class. They will be described below in separate subsections.

A. *LINEAR COORDINATE TRANSFORMATIONS*

In this subsection it will be shown that if ϕ represents a linear coordinate transformation,

$$\underline{x} = \phi\underline{z}, \tag{55}$$

then the five coordinate selections evaluated in Section V, when obtained from further transformations of $\underline{z}$, yield MEIs equal to those obtained from the equivalent coordinate selections based on transforming $\underline{x}$. To show this, consider coordinate transformations T of (17) and T_2 outlined in Fig. 1 and defined again below:

$$\underline{z} = T_2\underline{\xi}. \tag{56}$$

The argument is similar for any of the coordinate selections discussed in Section V. As a generic example, suppose T and T_2 are chosen to produce dynamically uncoupled coordinates $\underline{\eta}$ and $\underline{\xi}$. Combining (55) and (56)

$$\underline{x} = \phi T_2\underline{\xi} \triangleq T_3\underline{\xi}. \tag{57}$$

The relationship between the state matrices in $\underline{\xi}$ and $\underline{x}$ coordinates can be obtained as in (18):

$$\bar{A}_\xi = T_3^{-1}AT_3; \tag{58}$$

and directly from (18) the state matrix in $\underline{\eta}$ coordinates is

$$A = T^{-1}AT. \tag{59}$$

For this article we will assume that A_ξ and $\bar{A}$ are diagonal. (The block diagonal case is slightly more complicated and will not be presented here.)

For the diagonal case, T and T_3 are modal matrices of the eigenvectors of A. Thus, they are essentially the same except for a possible scaling of the eigenvectors. Such a possible scaling can be represented by a diagonal matrix S:

$$T = ST_3. \tag{60}$$

From (17), (57), and (60),

$$\underline{\eta} = s\underline{\xi}. \tag{61}$$

Such a diagonal coordinate transformation as (61) was called a "skewing" in Section VI, where it was shown that

$$MEI_\eta = MEI_\xi. \tag{62}$$

Therefore, the quality of CCA model reduction in dynamically uncoupled coordinates based on $\underline{x}$ is the same as the equivalent truncation based on $\underline{z}$. In fact, the dynamically uncoupled coordinates obtained from either $\underline{x}$ or $\underline{z}$ are essentially the same, since the dynamically uncoupled coordinates of a system in form (1) are invariant under coordinate transformations of type (55).

The argument is the same for any of the other coordinate selections mentioned in Section V, and it will not be repeated. Since the MEI_{η} is the same MEI_{ξ}, it is permissible to perform transformation (55) and then evaluate coordinate selections based on $\underline{z}$ instead of $\underline{x}$. If $\underline{z}$ is chosen judiciously, the system in (25) can have fewer parameters than the system in (1), making the evaluation of coordinate selections more practical.

It may appear intuitive that a particular set of uncoupled coordinates based on $\underline{z}$ is essentially the same as the uncoupled coordinates $\underline{\eta}$ based directly on $\underline{x}$. In other words, the different uncoupled coordinates are unique to the system in the same way as the eigenvalues are unchanged by intermediate coordinate transformations.

The system changes described in the following subsections are not coordinate transformations, and it is less intuitive to show that they are permissible.

B. TIME SCALING

Consider the time scaling

$$\tau = \alpha t. \tag{63}$$

This implies

$$\dot{\underline{x}} = \alpha \, d\underline{x}/d\tau. \tag{64}$$

In terms of τ, system (1) becomes

$$d\underline{\tilde{x}}/d\tau = \tilde{A}\underline{\tilde{x}} + \tilde{D}\underline{w}, \quad \tilde{A} = (1/\alpha)A,$$
$$\underline{y} = \tilde{C}\underline{\tilde{x}}, \quad \tilde{D} = (1/\alpha)D, \quad \tilde{C} = C. \tag{65}$$

It will be shown that the time-scaling change from system (1) to system (65) does not affect the evaluation of coordinate selections for CCA model reduction. The tildes used to denote time-scaled variables should not be confused with the skewed variables of Section VI.

Consider the MEIs of two dynamically uncoupled coordinates: $\underline{\eta}$ as obtained by (17) from the system in $\underline{x}$ and $\underline{\theta}$ obtained from the time-scaled coordinates:

$$\underline{\tilde{x}} = \tilde{T}\theta. \tag{66}$$

In θ coordinates, the system of (65) becomes

$$d\underline{\theta}/d\tau = A_\theta\underline{\theta} + D_\theta\underline{w}, \quad A_\theta = \tilde{T}^{-1}\tilde{A}\tilde{T},$$
$$\underline{y} = C_\theta\underline{\theta}, \quad D_\theta = \tilde{T}^{-1}\tilde{D}, \quad C_\theta = \tilde{C}\tilde{T}. \tag{67}$$

Since $\underline{\eta}$ and $\underline{\theta}$ are the dynamically uncoupled coordinates obtained from $\underline{x}$ and $\underline{\tilde{x}}$, the columns of T and $\tilde{T}$ are the eigenvectors of the following matrix eigensystem equations:

$$[A - \lambda I]\underline{t} = \underline{0}, \quad [\tilde{A} - \tilde{\lambda}I]\underline{\tilde{t}} = \underline{0}. \tag{68}$$

Using (65), the second equation of (68) is

$$[A - \alpha\tilde{\lambda}I]\underline{\tilde{t}} = \underline{0}. \tag{69}$$

From (68) and (69) it is clear that $T = \tilde{T}$. Therefore from (18) and (65),

$$A_\theta = (1/\alpha)\overline{A}, \quad D_\theta = (1/\alpha)\overline{D}, \quad C_\theta = \overline{C}. \tag{70}$$

$\overline{K}$ is obtained from (20). K_θ is obtained from

$$(1/\alpha)\overline{A}^T K + (1/\alpha)K\overline{A} + \overline{C}^T\overline{C} = 0. \tag{71}$$

It is observed that

$$K_\theta = \alpha\overline{K}. \tag{72}$$

Directly from (70) and the definition of G in (20),

$$G_\theta = (1/\alpha^2)\overline{G}. \tag{73}$$

The "cost" V of (2) in the two systems is related by

$$V_\theta = \mathrm{tr}\ K_\theta G_\theta = (1/\alpha)\ \mathrm{tr}\ \overline{K}\overline{G} = (1/\alpha)\overline{V}. \tag{74}$$

To compute the MEIs, the primed systems are formed in $\underline{\eta}$ and $\underline{\theta}$. As seen from (15), A', D', and C' are formed from the elements of A, D, and C. Therefore,

$$A'_\theta = (1/\alpha)\overline{A}' \tag{75}$$

because for all i, j $a_\Theta(i, j) = (1/\alpha)\overline{a}(i, j)$. Similarly,

$$D'_\theta = (1/\alpha)D', \quad C'_\theta = C'. \tag{76}$$

δV is the cost of the primed system (16) in the same way as V is the cost of the original system. Since (70) implies (74), (75) and (76) imply

$$\delta V_\theta = (1/\alpha)\ \delta V. \tag{77}$$

From (14), (74), and (77),

$$\mathrm{MEI}_\theta = \mathrm{MEI}_\eta. \tag{78}$$

So far it has been shown that the quality of CCA model reduction in the dynamically uncoupled coordinates obtained from $\underline{x}$ is the same as the quality of CCA model reduction in the dynamically uncoupled coordinates obtained from the time-scaled coordinates $\underline{\tilde{x}}$. An identical sequence of arguments, beginning with (68), can be carried out for the other four coordinate selections listed in Section V. Therefore, the evaluation of coordinate selections is unbiased and time scaling is a permissible system transformation.

C. INPUT, OUTPUT, AND STATE SCALING

In this section it will be demonstrated that it is permissible to simply multiply any of the system matrices of (1) by a scalar. For example, consider the input scaling

$$D_u = \alpha D, \tag{79}$$

which changes system (1) into

$$\dot{\underline{u}} = A\underline{u} + D_u\underline{w}, \quad \underline{y} = C\underline{u}. \tag{80}$$

It can be shown that $MEI_u = MEI_x$. MOre importantly, it can be shown that $MEI_{\tilde{u}} = MEI_{\tilde{x}}$, where $\underline{\tilde{u}}$ and $\underline{\tilde{x}}$ are equivalent coordinates obtained from $\underline{u}$ and $\underline{x}$. The argument proceeds in much the same way as in Section VIIB and is very straight forward. The matrices that participate in $V_{\tilde{u}}$ and $V_{\tilde{x}}$ are calculated, and the relation is

$$V_{\tilde{u}} = \alpha^2 V_{\tilde{x}}. \tag{81}$$

Similarly, evaluating the matrices composing $\delta V_{\tilde{u}}$ and $\delta V_{\tilde{x}}$ shows that

$$\delta V_{\tilde{u}} = \alpha^2 \ \delta V_{\tilde{x}}, \tag{82}$$

and the result is proven.

In the same way, it can be proven that output scaling or state scaling [i.e., the multiplication of C or A in (1) by a scalar] does not bias the evaluation of coordinate selections for use in CCA model reduction.

It may be well to state explicitly that the input vector $\underline{w}$ can be rewritten in terms of another input vector v of covariance V:

$$\underline{w} = M\underline{v}, \quad W = MVM^T. \tag{83}$$

System (1) becomes

$$\underline{\dot{x}} = A\underline{x} + D_V\underline{v}, \quad D_V = DM,$$
$$\underline{y} = C\underline{x}, \quad E\left[\underline{v}(t)\underline{v}^T(\tau)\right] = V\,\delta(t - \tau). \tag{84}$$

The input matrix G is the same as in (4):

$$G = D_V V D_V^T = DWD^T. \tag{85}$$

The state and output matrices are also unchanged in the new system. Therefore, CCA model reduction based on any coordinates derived from $\underline{x}$ in (84) is identical to the equivalent reduction based on (1), since the two systems are in fact the same.

VIII. PARAMETER REDUCTION

The "permissible" transformations of Section VII will be applied here to reduce the system defined in (1) to one with fewer parameters. The system of (1) is shown here for convenience:

$$\dot{x} = A\underline{x} + D\underline{w}, \quad E[\underline{w}(t)] = 0,$$
$$\underline{y} = C\underline{x}, \quad E\left[\underline{w}(t)\underline{w}^T(\tau)\right] = W\,\delta(t - \tau). \tag{86}$$

The number of parameters present in it is

$$N_{86} = n^2 + nk + np + k(k + 1)/2. \tag{87}$$

First, system (86) will be changed by a coordinate transformation to a dynamically uncoupled system:

$$\underline{\dot{\eta}} = \Lambda\underline{\eta} + Dw,$$
$$\underline{y} = C\underline{\eta}, \quad E\left[\underline{w}\ \ \underline{w}^T\right] = W. \tag{88}$$

In what follows it will be assumed that the Λ matrix is diagonal. System (88) contains

$$N_{88} = n + nk + np + k(k + 1)/2 \tag{89}$$

parameters. Second, the change in input vector of Section VII,C is applied to (88). Let

$$W = MM^T, \quad \underline{w} = M\underline{v}. \tag{90}$$

Then

$$\dot{\underline{\eta}} = \Lambda\underline{\eta} + \overline{D}_v\underline{v}, \quad \overline{D}_v = DM,$$
$$y = \overline{C}\underline{\eta}, \quad E[\underline{v}\ \ \underline{v}^T] = V = I. \tag{91}$$

The input covariance matrix is unity. Thus all the $k(k + 1)/2$ parameters of W have been eliminated in (91), and their effect is absorbed in D_v. The number of remaining parameters is

$$N_{91} = n + nk + np. \tag{92}$$

Third, another coordinate transformation,

$$\underline{\eta} = \psi\underline{z}, \tag{93}$$

is performed, with

$$\psi(i, j) = 0 \quad \text{for} \quad i \neq j,$$
$$\psi(i, j) = 1/c(i, j). \tag{94}$$

System (91) becomes

$$\underline{z} = \Lambda\underline{z} + \hat{D}\underline{v}, \quad \hat{D} = \psi^{-1}\overline{D}_v,$$
$$\underline{y} = \hat{C}z, \quad \hat{C} = \overline{C}\psi. \tag{95}$$

Λ is the same as in (88) because it is diagonal. The first row in C contains all ones. Therefore, n more parameters have been eliminated, and

$$N_{95} = nk + np. \tag{96}$$

The case where any of the $C_{ij} = 0$ requires a slightly different approach that will not be addressed here.

Fourth, a time scaling is implemented:

$$\tau = -\lambda_1 t, \tag{97}$$

where λ_1 is the first eigenvalue of Λ. If the first eigenvalue of Λ is complex, the $\underline{z}$ vector can be reordered so that a real eigenvalue appears in the first entry of Λ. If there are no real eigenvalues, then λ_1 is taken as the real part of the first (complex) eigenvalue. After time scaling (95) becomes

$$d\tilde{z}/d\tau = \Lambda\underline{z} + D\underline{v}, \quad \Lambda = (1/\lambda_1)\Lambda,$$
$$\underline{y} = \tilde{C}\tilde{\underline{z}}, \quad \tilde{D} = (1/\lambda_1)D, \quad \tilde{C} = C. \tag{98}$$

The first entry in Λ is -1 (assuming stable A), and one more coefficient has been eliminated:

$$N_{98} = np + nk - 1. \tag{99}$$

If there are no real eigenvalues, the real parts of the first and second (i.e., conjugate) eigenvalues are one, so the number of parameters is also reduced by one.

Fifth, an input scaling (Section VII,C) is performed,

$$\tilde{B} = (1/d_{11})\tilde{D}, \tag{100}$$

and system (99) is replaced with

$$d\tilde{\underline{x}}/d\tau = \tilde{\Lambda}\tilde{\underline{x}} + \tilde{B}\underline{v}, \quad \underline{y} = \tilde{C}\tilde{\underline{x}}. \tag{101}$$

Clearly $b_{11} = 1$, so one more parameter has been eliminated:

$$N_{101} = np + nk - 2. \tag{102}$$

It may seem like the last two steps above are inconsequential when compared with the previous step, because they only reduce the number of parameters by one. This can still be very important in applications where n is small, as the example of Section IX will demonstrate.

One may wonder if N_{101} is the minimum number of parameters obtainable. Might not some other permissible transformations exist that reduce N_{101} further? This question is not fully answered, and the topic requires further investigation.

Since all the parameter reductions resulted from permissible transformations, the evaluation of coordinate selections for CCA model reduction based on (101) is the same as based on (86). As the example of the next section will show, it may be more convenient (or even necessary) to perform the comparative evaluation based on the system of (101).

IX. EXHAUSTIVE EVALUATION OF COORDINATE SELECTIONS

The two-coordinate SISO case used in the numerical example of Section V led to the question, In which regions in the parameter space do various coordinate selections perform best? With the help of parameter simplication an exhaustive search becomes possible.

In parametric form, the two-dimensional SISO example of (86) is

$$\begin{Bmatrix} \dot{x}_1 \\ \dot{x}_2 \end{Bmatrix} = \begin{bmatrix} a_{11} & a_{12} \\ a_{21} & a_{22} \end{bmatrix} \begin{Bmatrix} x_1 \\ x_2 \end{Bmatrix} + \begin{bmatrix} d_1 \\ d_2 \end{bmatrix} w,$$

$$y = [c_1 \quad c_2] \begin{Bmatrix} x_1 \\ x_2 \end{Bmatrix}, \qquad E[w^2] = \sigma^2. \tag{103}$$

The number of parameters present in (103) can be counted as nine, which verifies N_{86} in (87) with $n = 2$, $p = 1$, and $k = 1$. The five parameter-simplifying steps discussed in Sections VII and VIII are applied to (103) to obtain the equivalent of (101):

$$\begin{Bmatrix} \dot{\tilde{x}}_1 \\ \dot{\tilde{x}}_2 \end{Bmatrix} = \begin{bmatrix} -1 & 0 \\ 0 & \tilde{\lambda} \end{bmatrix} \begin{Bmatrix} x_1 \\ x_2 \end{Bmatrix} + \begin{bmatrix} 1 \\ \tilde{b} \end{bmatrix} v$$

$$y = [1 \quad 1]\begin{Bmatrix} \tilde{x}_1 \\ \tilde{x}_2 \end{Bmatrix}; \qquad E[v^2] = 1$$

Seven parameters are eliminated in this case.

The model reduction involved here is the simplest possible: from a two-coordinate to a single-coordinate system. Even so, it is not possible to analytically determine which coordinate selection is best for performing CCA mdel reduction. For this reason, a thorough numerical evaluation of CCA truncation in the various coordinate selections is considered desirable (104). To this end, iteration on the two parameters left in is more convenient than a similar iteration upon the nine (103). A computer search in the two-dimensional parameter space of (104) can be both broad enough and dense enough to discover all trends, and the results can be displayed graphically. The nine-dimensional parameter space of (103) makes any thorough investigation impractical.

The parameter space of interest is for all real $\tilde{b}$ and real negative $\tilde{\lambda}$ (stable state matrix with real eigenvalues assumed). Furthermore, $\tilde{\lambda} > -1$ is not of interest, because for any system (64) with $\tilde{\lambda} < -1$ there is an equivalent system with $\tilde{\lambda} > -1$ that contains the identical CCA truncation information [1]. Therefore, a complete search through the space shown in Fig. 2 is

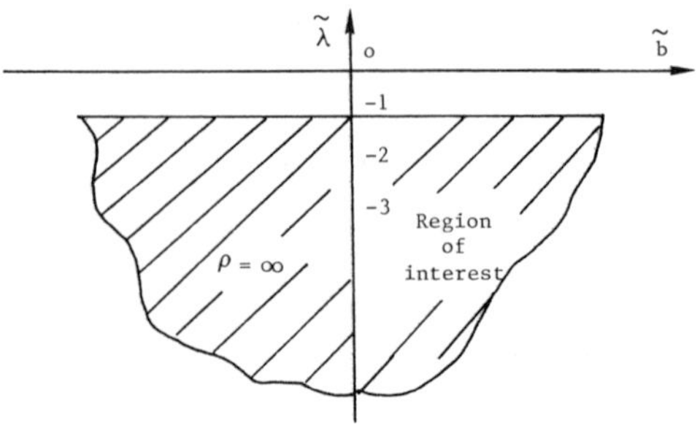

Fig. 2. Search space.

performed. It was found that values of b from -500 to 500 revealed all trends of interest, whereas λ had to be increased considerably ($\sim 10^{12}$) before a final pattern was established.

For each point in the search space, the five coordinate selections of Section V are obtained. The MEI for CCA model reduction in each of the five coordinates is then calculated. The results are presented graphically as a map of the region of interest, where the coordinate selection providing the lowest MEI is indicated in every area.

An overview of these results is shown in Fig. 3. As seen, the balanced coordinates produce the best quality CCA model reduction (as measured by a low MEI) most often. The reason the results are so detailed is because δV of (14) (and V also) is a complicated rational function of polynomials in the system parameters. Comments, interpretation of the results, and more

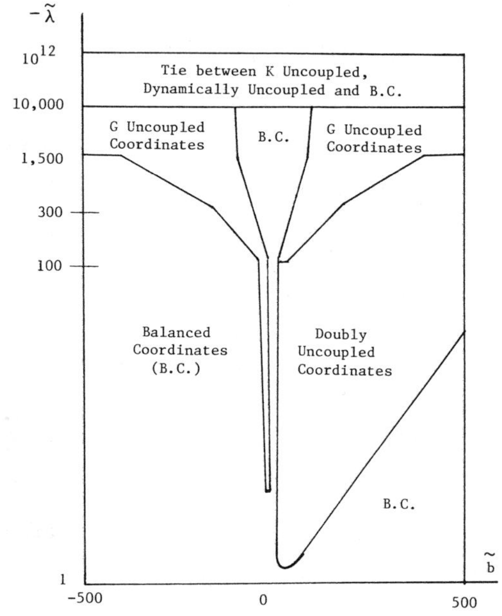

Fig. 3. Overview of low-MEI coordinate selections.

detailed graphical displays can be found in [1]. Although some particular points in the parameter space of Fig. 3 have their results explained in [1], the basic question of predicting the best coordinate selection for CCA model reduction remains. In the two-dimensional SISO example considered, it is possible to determine which coordinate selection to use by first parameter simplifying and then checking against the chart of Fig. 3. But in the general-dimensional MIMO case, the search for the holy coordinates remained a topic for further research.

X. SUMMARY

The field of research in the order reduction of linear systems is vast and growing. A brief survey was provided. Among the component truncation approaches, the method of cost decomposition provides an appealing framework for model reduction. It takes into account all three of the major influences on model fidelity: input coupling, output coupling, and frequency spectrum. The success of the method, however, is strongly affected by the particular coordinates selected for describing the (untruncated) large-order model.

This article examines the performance of this method for a particular class of coordinate selections; those that uncouple the cost function. The intuitive appeal of these coordinates is discussed. Several cost-uncoupling coordinate transformations are presented, including a "doubly cost uncoupling" approach that simultaneously diagonalizes the two-component matrices making up the cost function. In testing the approach on a large number of simple examples, it was found that (a) in the vast majority of cases (but not all), truncation of low-cost components leads to better quality reduced models than truncation of high-cost components and (b) doubly uncoupled coordinates or balanced coordinates are more likely to produce good quality reduced models than the other coordinate selections studied. It is important to recognize, however, that these are only trends. Thus, while doubly cost-uncoupled coordinates and balanced coordinates seem to perform better than the other three coordinate selection schemes, none of the selections considered was best in all cases, and each of the cost-uncoupling schemes was best in some cases. These results clearly indicate the need for further investigation in this area.

In addition to these numerical results, two analytical results were obtained concerning properties of the reduced models. First it was shown that if the large-model coordinates are uncoupled by virtue of K being uncoupled, then asymptotic stability of the large model ensures stability of the reduced model. Furthermore, it was shown that skewing of coordinate transformations can neither improve nor degrade the quality of the reduced model.

Finally, it was shown that several system changes are permissible for CCA model reduction (i.e., they do not bias the evaluation of coordinate selections). These system changes were applied in such a way as to reduce the number of parameters of a general linear state-space system. An example showed that application of parameter simplification was a key step toward a thorough evaluation of coordinate selections for CCA model reduction on a low-order system. In addition to the application at hand, the various steps used in generating systems containing fewer parameters can serve as a guideline for similar steps appropriate in different other problems.

APPENDIX: FURTHER NUMERICAL EXAMPLES

This appendix contains several examples whose results differ from the trends suggested by Table 1. The various cases are listed below, and their results are collected together in Table 2, which has the same format as Table 1 but does not include parameter optimal results.

Table 2. Further Comparison of Truncation Schemes

Coordinate selection	MEI for low-cost component truncation	MEI for high-cost component truncation
Case 1		
Original	0.9250	0.9891
G uncoupled	0.7557	0.5839[b]
K uncoupled	0.7667	0.5790[b]
Double uncoupled	1.0907	0.6300[b]
Balanced coordinates	0.7022	0.5867[b]
A uncoupled	0.9250	0.9891
Case 2		
Original	0.1625	0.7345
G uncoupled	0.1596[a]	0.9306
K uncoupled	0.1650	0.7027
Double uncoupled	0.1604	1.0562
Balanced coordinates	0.1691	0.6640
A uncoupled	0.1625	0.7345
Case 3		
Original	0.0571	0.7121
G uncoupled	0.2913	1.0001
K uncoupled	0.0227[a]	0.8069
Double uncoupled	0.0986	0.9999
Balanced coordinates	0.0229	0.8009
A uncoupled	0.0571	0.7121
Case 4		
Original	0.1238	0.8315
G uncoupled	0.1628	1.0265
K uncoupled	0.0688	0.9675
Double uncoupled	0.0737	0.9869
Balanced coordinates	0.0594[a]	0.8799
A uncoupled	0.1238	0.8315

[a] *Identifies the lowest MEI.*

[b] *Identifies cases where high-cost component truncation is best.*

Case 1. This is an example in which truncation of the high-cost component is better than truncation of the low-cost component in G-uncoupled, K-uncoupled, double-uncoupled, and balanced coordinates (see also Case 4). The system is

$$\dot{x} = \begin{bmatrix} -1 & 0 \\ 0 & -7.5 \end{bmatrix} x + \begin{bmatrix} 1 & 1 \\ 20 & 1 \end{bmatrix} w, \quad E[w(t)] = 0,$$

$$y = [1 \quad -0.2]x, \quad E[w(t)w^T(\tau)] = \delta I(t - \tau).$$

Case 2. In this case, the G-uncoupled coordinate selection produces the lowest MEI:

$$x = \begin{bmatrix} -1 & 0 \\ 0 & -90 \end{bmatrix} x + \begin{bmatrix} 1 & 1 \\ 30 & 2 \end{bmatrix} w, \quad E[w(t) = 0,$$

$$y = (1 \quad 1)x, \quad E[w(t)w^T(\tau)] = I\ \delta(t - \tau).$$

Case 3. In this case, the K-uncoupled coordinate selection produces the lowest MEI:

$$\dot{x} = \begin{bmatrix} -1 & 0 \\ 0 & 10 \end{bmatrix} x + \begin{bmatrix} 1 & 0.1 \\ 0.9 & 0 \end{bmatrix} w, \quad E[w(t)] = 0,$$

$$y = (1 \quad 1)x, \quad E[w(t)w^T(\tau) = I\ \delta(t - \tau).$$

Case 4. Here, truncation in balanced coordinates is best. Note that the lowest MEIs are produced by truncation of the low-cost components (unlike Case 1).

$$\dot{x} = \begin{bmatrix} -1 & 0 \\ 0 & -6 \end{bmatrix} x + \begin{bmatrix} 1 & 0.1 \\ 0.9 & 0.5 \end{bmatrix} w, \quad E[w(t)] = 0,$$

$$y = (1 \quad 1)x, \quad E[w(t)w^T(\tau)] = I\ \delta(t - \tau).$$

REFERENCES

1. A. L. DORAN, "Model Reduction by Cost Decomposition: Implications of Coordinate Selection," Ph.D. Dissertation, University of California, Los Angeles, 1982.

2. R. E. SKELTON and P. C. HUGHES, "Modal Cost Analysis for Linear Matrix Second Order Systems," *J. Dyn. Syst. Meas. Control 102*, 151-158 (1980).

3. H. B. HABLANI, P. C. HUGHES, and R. E. SKELTON, "Generic Model of a Large Flexible Spacecraft," Technical Report 955369, JPL, 1980.

4. R. GRAN and M. ROSSI, "A Survey of the Large Structures Control Problem," *Proc. IEEE Conf. Decision Control*, 1002-1007 (1979).

5. W. F. POWERS, "The Dynamics and Control of Large Space Structures: An Overview," *J. Astronaut. Sci. 27*, No. 2, 95-101 (1979).

6. P. W. LIKINS, "The New Generation of Dynamic Interaction Problems," *J. Astronaut. Sci. 27*, No. 2, 103-113 (1979).

7. L. MEIER and D. G. LUENBERGER, "Approximation of Linear Constant Systems," *IEEE Trans. Autom. Control AC-12*, 585-588 (1967).

8. R. GENESIO and M. MILANESE, "A Note on the Differentiation and Use of Reduced Order Models," *IEEE Trans. Autom. Control*, 118-122 (1976).

9. M. J. BOSLEY and F. P. LEES, "A Survey of Simple Transfer Function Derivations from High Order State Variable Models," *Automatica 18*, 765-775 (1972).

10. R. H. ROSSEN and L. LAPIDUS, "Minimum Realizations and System Modeling 1. Fundamental Theory and Algorithms," *AICE J. 18*, No. 4, 673-684 (1972).

11. N. R. SANDELL, JR., P. VARAIYA, and M. ATHANS, "A Survey of Decentralized Control Methods for Large Scale Systems, 1975 Systems Engineering for Power: Status and Prospects, *U.S. Energy Dev. Adm.*, 334 (1975).

12. D. M. FELLOWS, N. K. SINHA, and J. C. WISMATH, "Reductions of the Order of Dynamic Models," *Natl. Conf. Autom. Control (Canada)* (1970).

13. D. A. WILSON, "Optimum Solution of Model Reduction Problem," *Proc. IEEE 177*, No. 6, 1161-1165 (1970).

14. C. F. CHEN, "Model Reduction of Multivariable Control Systems by Means of Matrix Continued Fractions," *Int. J. Control 10*, No. 2, 225-238 (1974).

15. J. B. RIGGS and T. F. EDGAR, "Least Squares Reduction of Linear Systems Using Impulse Response," *Int. J. Control 20*, No. 2, 213-223 (1974).

16. R. E. SKELTON, "Cost Decomposition of Linear Systems with Application to Model Reduction," *Int. J. Control 32*, No. 6, 1031-1055 (1980).

17. R. E. SKELTON and A. YOUSUFF, "Component Cost Analysis of Large Scale Systems," "Control and Dynamic Systems," Vol. 18, pp. 1-54 (C. T. Leondes, ed.), Academic Press, New York, 1982.

18. B. C. MOORE, "Principal Component Analysis in Linear Systems: Controllability, Observability and Model Reduction," *IEEE Trans. Autom. Control AC-26*, No. 1, 17-32 (1981).

19. L. PARNEBO and L. M. SILVERMAN, "Model Reduction via Balanced State Space Representative," *IEEE Trans. Autom. Control AC-27*, 382-387 (1982).

20. D. C. HYLAND, "The Optimal Projection Approach to Fixed-Order Compensation: Numerical Methods and Illustrative Results," *AIAA 21st Aerosp. Sci. Meet., Reno, Nevada* (1983).

21. D. C. HYLAND, "Comparison of Various Controller-Reduction Methods: Suboptimal versus Optimal Projection," *AIAA Dyn. Spec. Conf., Palm Springs* (1984).

22. D. C. HYLAND and D. S. BERNSTEIN, "The Optimal Projection Approach to Model Reduction and the Relationship between the Methods of Wilson and Moore," *Proc. 23rd IEEE Conf. Decision Control, Las Vegas* (1984).

23. L. G. GIBILARO and F. P. LEES, "The Reduction of Complex transfer Function Models to Simple Models by the Method of Moments," *Chem. Eng. Sci. 24*, 85-93 (1969).

24. F. P. LEES, "The Derivation of Simple Transfer Function Models of Oscillatory and Inverting Processes from the Basic Transformed Equations Using the Method of Moments," *Chem. Eng. Sci. 26*, 1179-1186 (1971).

25. H. W. KROPHOLLER, "The Determination of Relative Variance and Other Moments for Generalized Flow Networks or System Transfer Functions," *Ind. Eng. Chem. Fundam. 9*, 329-333 (1970).

26. V. ZAKIAN, "Simplication of Linear Time Invariance Systems by Moment Approximants," *Int. J. Control 18*, No. 3, 455-460 (1973).

27. Y. SHAMASH, "Linear System Reduction Using Padé Approximation to Allow Retention of Dominant Modes," *Int. J. Control 21*, No. 2, 257-272 (1975).

28. R. K. APPIAH, "Padé Methods of Horwitz Polynomial Approximation with Application to Linear System Reduction," *Int. J. Control 29*, No. 1, 39-48 (1979).

29. Y. SHAMASH, "Order Reduction of Linear Systems by Padé Approximation Methods," Ph.D. Thesis, Imperial College of Science and Technology, England, 1973.

30. Y. SHAMASH, "Stable Based Reduced Order Models Using the Routh Technique," *Proc. IEEE Conf. Decision Control*, 864 (1979).

31. K. C. DALY and A. P. COLEBOURN, "Padé Approximation for State Space Models," *Int. J. Control 30*, No. 1, 37-47 (1979).

32. C. F. CHEN and L. S. SHIEH, "A Novel Approach to Linear Model Simplification," *Int. J. Control 8*, No. 6, 561-570 (1968).

33. S. C. CHUANG, "Application of Continued Fraction Method for Modeling Transfer Functions to Give More Accurate Initial Transient Response," *Electron. Lett. 6*, No. 26, 861-863 (1970).

34. D. J. WRIGHT, "The Continued Fraction Representation of Transfer Functions and Model Simplification," *Int. J. Control 18*, No. 3, 449-454 (1973).

35. C. T. CHEN, "A Formula and Algorithm for Continued Fraction Inversion," *Proc. IEEE 57*, 1780-1781 (1969).

36. C. F. CHEN, C. J. CHUANG, and L. C. SHIEH, "Simple Methods for Identifying Linear Systems from Frequency of Time Response Data," *Int. J. Control 13*, 1027-1039 (1971).

37. H. S. WALL, "Analytic Theory of Continued Fractions," Chelsea, Bronx, New York, 1948.

38. N. K. SINHA and W. PILLE, "A New Method for Reduction of Dynamic Systems," *Int. J. Control 14*, No. 1, 111-118 (1971).

39. E. C. LEVY, "Complex Curve Fitting," *IRE Trans. Autom. Control 4*, 37-43 (1959).

40. N. K. SINHA and G. T. BEREZNAI, "Optimum Approximation of High Order Systems by Low Order Models," *Int. J. Control 14*, No. 5, 951-959 (1971).

41. E. J. DAVISON, "A Method for Simplifying Linear Dyanmic Systems," *IEEE Trans. Autom. Control AC-11*, 93-101 (1966).

42. E. J. DAVISON, "A New Method for Simplifying Large Linear Dynamic Systems," *IEEE Trans. Autom. Control AC-13*, 214-215 (1968).

43. M. R. CHIDAMBARA, "Two Simple Techniques for the Simplification of Large Dynamic Systems," *Proc. JACC*, 6669-6674 (1969).

44. R. NAGARAJAN, "Optimum Reduction of Large Dynamic System," *Int. J. Control 14*, No. 6, 1169-1174 (1971).

45. A. KUPPURAJULU and S. ELANGOVAN, "System Analysis by Simplified Models," *IEEE Trans. Autom. Control AC-15*, 234-237 (1970).

46. D. MITRA, "Analytic Results on the Use of Reduced Models in the Control of Linear Dynamical Systems," *Proc. IEEE 116*, 1439-1444 (1969).

47. D. MITRA, "The Reduction of Complexity of Linear Time Invariant Dynamic Systems," *Proc. IFAC Tech. Ser. 4th, Warsaw, 67*, 19-33 (1969).

48. S. A. MARSHALL, "An Approximate Method for Reducing the Order of a Linear System," *Control 10*, 642-643 (1966).

49. P. V. KOKOTOVIC, "A Control Engineer's Introduction to Singular Perturbations: Order Reduction in Control System Design," pp. 1-12, ASME Publ., New York, 1972.

50. P. V. KOKOTOVIC and P. SANUTI, "Singular Perturbations Method for Reducing the Model Order in Optimal Control Design," *IEEE Trans. Autom. Control AC-13*, No. 4 (1968).

51. P. V. KOKOTOVIC, R. E. O'MALLEY, and P. SANUTI, "Singular Perturbations and Order Reduction in Control Theory—An Overview," *Automatica 12*, 123-132 (1976).

52. K. V. KRIKORIAN and C. L. LEONDES, "Applications of Singular Perturbations to Optimal Control," *in* "Control and Dynamic Systems," Vol. 18, pp. 131-160 (C. T. Leondes, ed.), Academic Press, New York, 1982.

53. R. E. KALMAN, "Mathematical Description of Linear Dynamical Systems," *SIAM J. Control 1*, 152 (1963).

54. R. E. KALMAN, "Irreducible Realizations and the Degree of a Rational Matrix," *SIAM J. 13*, 520 (1965).

55. W. A. WOLOVICH and P. L. FALB, "On the Structure of Multivariable Systems," *SIAM J. Control 7*, 246 (1969).

56. M. AOKI, "Control of Large Scale Dynamic Systems by Aggregation," *IEEE Trans. Autom. Control AC-13*, No. 3, 246-253 (1968).

57. D. A. WILSON and R. N. MISHRA, "Optimal Reduction of Multivariable Systems," *Int. J. Control 29*, No. 2, 267-278 (1979).

58. D. A. WILSON, "Model Reduction for Multivariable Systems," *Int. J. Control 20*, No. 1, 57-64 (1974).

59. J. H. ANDERSON, "Geometrical Approach to the Reduction of Dynamical Systems," *Proc. IEEE 114*, No. 7 (1967).

60. G. HIRZINGER and G. KREISSELMEIER, "On Optimal Approximation of High Order Linear Systems by Low Order Models," *Int. J. Control 22*, No. 3, 399-408 (1975).

61. F. D. GALIANA, "On the Approximation of Multiple Input Multiple Output Constant Linear Systems," *Int. J. Control 17*, No. 6, 1313-1324 (1973).

62. H. KWAKERNAAK and R. SIVAN, "Linear Optimal Control Systems, Wiley (Interscience), New York, 1972.

63. Y. YAM, T. L. JOHNSON, and J. C. LIN, "Aggregation of Large Space Structure Dynamics with Respect to Actuator and Sensor Influences," *Proc. 20th IEEE Conf. Decision Control*, 936-942 (1981).

INDEX

I

K

L

M

N

O

P

Q

R